1. Newburg Pond
2. Lib. ...
3. Pratt - Broo...
4. Coast ...

ANAL

AILROADS

COAL BOATS TO TIDEWATER

To Austin M. Kellam,
with best personal regards
for a pleasant ride
down the old D. & H. Canal,

Manville B. Wakefield

COAL BOATS TO TIDEWATER

The Story of the
DELAWARE & HUDSON CANAL

by
Manville B. Wakefield

Foreword by
Carl Carmer

LIBRARY OF CONGRESS CATALOG

CARD NO. 65-27932

Printed by

Steingart associates, inc.

SOUTH FALLSBURG, NEW YORK

To

Barb

Debby, Lisa and Andy

Contents

← *"Narrows of the Lackawaxen" by M. B. Wakefield*

Maps

← ***Paymaster's Boat passing the S. A. VanWagner canal store in Ellenville***

M. B. WAKEFIELD

Maps *Con't.*

All maps, diagrams and line drawings by M. B. Wakefield

Diagrams

← ***The Tontine Coffee House (left) where in 1825 the Wurts brothers demonstrated the burning qualities of their coal.***

a word about the MAPS

The maps are based on Official Maps of the Delaware and Hudson Canal surveyed in 1854, drawn in 1856, and revised in 1865 by R. F. Lord, Jr.

The map content is divided into two types: the broad divisional or SECTIONAL maps of which there are ten, and the highly detailed village and small area maps of which there are twenty-four. Referral to end-sheet maps will indicate the page on which the large SECTIONAL maps may be quickly found. The SECTIONAL maps also indicate by an *asterisk* what towns and areas are shown in further detail in that particular chapter.

The seemingly incorrect spelling on the original maps has been faithfully followed with Hugernot instead of Huguenot as it is spelled today, Port Hyxon instead of Port Hickson, VanTuyle instead of VanTyle, and so on.

After the building of the aqueduct at High Falls and the re-alignment of the canal, the locks between that point and the Delaware Aqueduct were renumbered. On the maps the new lock numbering is noted in large numerals while the smaller number indicates the old.

The large hollow arrow denotes reference to a picture in the chapter which is identified by a prefix letter, L for Lackawaxen Section or N for Neversink Section, followed by the number as it occurs in the chapter text, (for example N 1 or L 6). It may happen that the same number will appear on overlapping maps such as the Lackawaxen Section and the Delaware and Lackawaxen Aqueduct chapter. The directional point of the arrow in most cases indicates the approximate angle of the camera.

The hollow arrow pointing to the break in the lines at extreme left and right hand edges of the SECTIONAL maps indicates the approximate overlap points.

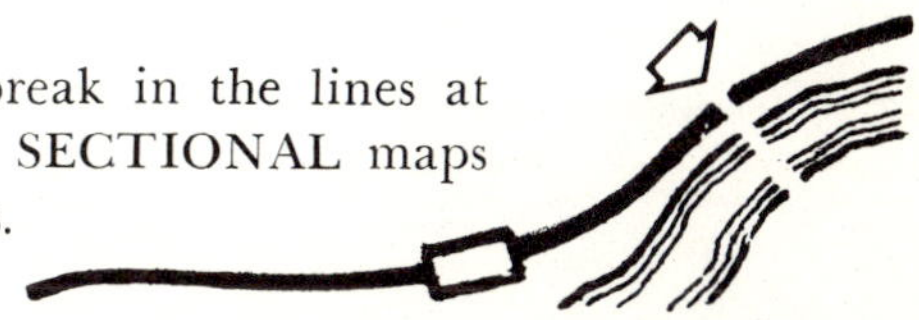

The popular name of the lock is shown directly below the lock number. Various lists were collated which indicated the names most commonly used by boatmen in the last twenty-five years of operation. Assistance to this end was generously forthcoming from Henry Soete, William Darragh, James A. O'Boyle, and the late DuBois Weber and John A. McGinnis.

← ***Canal at the Hawk's Nest***

Foreword

The major achievement of this remarkable book was the attainment of its overall concept. I know of no other recent volume of local history so comprehensively and logically planned.

Perhaps one added reason for the excellence here presented lies in the multiple gifts of the author. Mr. Wakefield has been able to maintain a unity which has emanated from the fact that one man has reasoned, written and illustrated the whole work. When mechanical drawings were required to clarify technical descriptions they were forthcoming from the same skilled hand that produced translations of old and faded photographs into scratch-board works of artistic merit. When the text required documented history, the mind of a scholar directed the study which supplied it. But the most treasured of the book's qualities comes not entirely from meticulous attention to details but from a genuine interest in the life of a past period and a genuine affection for the countryside through which the old canal passed.

Not such a magnificent and enormous undertaking as the Erie, the Delaware and Hudson had many distinctions not shared with the larger and longer channel. Canals have always appealed to the American imagination. Mr. Wakefield has sought out and included many colorful and amusing anecdotes. Several of them are documented, others, of a folk-character, have been told to him by "old-timers". The man-made waterway, wherever it occurs in our history, has been the subject of many a humorous gibe on the "sea-going" nature of its narrow, shallow channel.

This book is not merely a compilation of factual records and semi-literary episodes, however. It is a volume so compact and effective as to move me to recommend it to the attention of organizations which annually distribute recognitions of especial merit. It is also very obviously a labor of love. Mr. Wakefield imparts his own enthusiasm for the landscape which encompasses the old passage in every paragraph. Here and there he brings to his readers nostalgia for old days and ways, and he proves himself ever aware of horizons notched by the Shawangunk Mountains, of shimmering spring-fed pools, and of wooded areas unchanged and unchanging throughout the years.

Octagon House
Irvington-on-Hudson
New York
August, 1965

CARL CARMER

← ***Poling boats through Honesdale Basin***

M.B. WAKEF

Prologue

I remember

the muffled sound of horses straining at the whiffle trees, then the Canawler's cry "Look for a post."

I remember

the thick resined ropes sliding over my stone flanks, braced and anchored deep in the shoulder of the lock. Coal boats to tidewater, flicker or log-bilge, squeezer or futtock, snubbing them all alike.

I remember

the voices of boatmen, high in anger or pleasant and friendly as a warm August day, slapping at my flanks, with the water rising and falling.

I remember

the lock-keep in summer rush, too busy to leave, munching out of hand on pritties and skimmagig, or snubbing boat after boat, wearing my scars of service deeper and shinier.

I remember

the cries of the boatmen coming farther and farther apart - - - the ponderous gates creaking open less often - - - the push of coal boats to tidewater coming slower and slower as the quiet settled down. The wind flapping the shutters on the deserted lock-keepers house.

← ***Bolton Basin of* 1885**

The turbulent waters splashing endlessly below the towpath. In the distance I could hear the wail of a locomotive on a new born railroad and later still the roar of traffic on a concrete highway.

I remember

moss growing on the lock. Muskrats burrowing in the crumbling banks and the weeds creeping inexorably down. Cat-tails and loose strife, briar and sumac pushed in where once the trim gates lowered and raised the boats loaded with fuel for the hearth fires of the cities; muscle and brawn for the belching foundries and factories of a young and growing country.

Powerless, at a drunken angle in the settling ground, cast aside by a new age, I waited for decades.

One day I felt a knowing and sympathetic hand upon my deepest scars.

After oblivion someone had discovered me.

The discovery was mutual.

It was in the middle of a pleasant sunny afternoon in early April of 1965 that I had discovered the snubbing post.

Rope burns so shiny that the glint from the sunlight caught my eye before the scar-worn flanks of the post took shape.

The smooth patina of those indelible marks that years of weathering could not obliterate gave drama and immediacy to the story of the canal.

This forlorn and lonely snubbing post symbolized for me the fascination of a leisurely but dynamic way of life and the reason for this book.

To this post I hitch my story of Coal Boats To Tidewater.

The Beginnings

In 1825 the three hundred and sixty-three mile Erie Canal was completed. In the same year, the Schuylkill Canal, which ran from Mount Carbon to Philadelphia, was opened for business.

These two engineering achievements precipitated an economic malady that was soon to be diagnosed as *canal fever.* Later the railroads would create the same type of contagion but in the early 1820's it was, "Let's build a canal."

America was growing and a means of cheap mass transportation was needed. Canals were the answer. New York with a population of 202,000 and Philadelphia boasting 130,000 were both establishing themselves as centers of commerce and industry and needed certain production fuels. One import they were sorely in need of was bituminous coal, brought in from Europe before the War of 1812, but now in the post-war period in very short supply.

With this shortage existing, Philadelphians turned to their native hills of Pennsylvania for this vital industrial commodity. One of the first local suppliers was the Lehigh Company which in 1820 shipped in three hundred and sixty-five tons.

To go back a bit, in 1796, Obadiah Gore, a blacksmith who emigrated from Connecticut, used anthracite in his shop near Wilkes-Barre. In 1788 Jesse Fell of Wilkes-Barre, a blacksmith and later a judge, used anthracite in the manufacture of nails and by 1808 was using it for heating in a common grate in his fireplace.

The WURTS BROTHERS

The Swiss brothers, Maurice and William Wurts, [1] were engaged in a lucrative dry goods business in Philadelphia. During their years of business dealings they contracted a number of bad debts which were covered by collateral consisting of tracts of forest land.

William Wurts was already an avid hunter and knew much of the hills and forests. He was also aware of the frequent appearance of a black stone that appeared in certain rock out-croppings in the Lackawanna Valley. Associations with hunters and trappers on hunting trips in the hills enlightened Wurts as to the potential value of the black rock.

In addition to the debt-acquired lands, other wild tracts were picked up by the brothers at bargain prices of from fifty cents to three dollars an acre. About 1816 a small quantity of anthracite was mined and laboriously transported to Philadelphia where they were unable to sell it.

All of these frustrating efforts brought the brothers to what was later to become known as Carbondale. In 1822 a log house and camp was established there for themselves and their workmen. During the winter of 1822-23 Maurice and William Wurts rafted down to Philadelphia slightly over 100 tons of coal to what they sadly discovered to be a market well supplied with Lehigh coal.

Courtesy of DELAWARE & HUDSON CO.

William Wurts, one of the founders of the company.

It soon became obvious to the brothers that if they were to sell their coal they would have to find some other market. A glance at a map indicated their only other possible mass market was New York City and it was in this direction they concentrated their efforts.

A mass market potential such as New York would require a means of cheap bulk-type transportation. The solution - construct a canal.

SURVEY

In May of 1823 the Wurts brothers engaged Benjamin Wright, then the principal engineer on the Erie Canal, to make a survey from the Hudson River to a point as near the coal mines at Carbondale as possible.

Maurice suggested to Wright that the valley of the Neversink, the swamp lands of the mid-valley saddle, and the Rondout valley be embraced in some feasible plan.

Their first proposal was to dig a canal from Kingston to Carpenters Point (Port Jervis). Then, by the use of nine dams across the Delaware and seventeen more on the Lackawaxen, slack water navigation on these rivers would be possible. In late 1825 the idea of dams on the two rivers was abandoned in favor of an independent canal system along the banks of the rivers. 2

The certainty of conflict with the raftsmen was one of the prime motivating factors behind this change of plan.

In January 1824, Wright, his assistant Sullivan, and Sullivan's assistant John B. Mills, finished their surveys and published their report. A canal four feet deep and thirty-two feet wide at the surface, they said, could be dug for $1,208,632.95. This included an extra 5 percent for unseen expenses. 3

Courtesy of DELAWARE & HUDSON CO.

Maurice Wurts, one of the founders of the company and an original manager, **1825 - 1826,** ***and*** **1848 - 1854.**

TONTINE COFFEE HOUSE

In December of 1824 the Wurts brothers rafted a load of coal to Philadelphia where it was transshipped to New York City aboard the sloop *Tripler*. There, on a cold January 7th, in 1825, a momentous demonstration took place. (Opposite, xi)

In a specially constructed grate at the Tontine Coffee House 4 a roaring *stone coal* fire was prepared. With the chilling winds off the harbor rattling the windows, the men of finance and means were impressed, to the extent that the stock books were over-subscribed by two o'clock in the afternoon. At the same time, stock books were being closed in the Middle District Branch Bank in Kingston and at the Orange County Bank in Goshen.

On March 8, 1825, again at the Tontine Coffee House, the first Board of Managers of the Delaware and Hudson Canal Company was elected. The canal project was now proceeding under the official title of *The President, Managers and Company of the Delaware and Hudson Canal Company* with the following directors elected: Garret B. Abeel, John Bolton, Lynde Catlin,

Courtesy of
DELAWARE & HUDSON CO.

***Benjamin Wright, first chief engineer of the D. and H. Canal Co.,* 1825 - 1827.**

Philip Hone, John Hunter, Abram Hasbrouck, Rufus L. Lord, Hezekiah B. Pierrepont, Benjamin W. Rogers, Jonathan Thompson, George D. Wickham, Maurice Wurts and Samuel Whittemore.

March 11, 1825 saw forty-four year old Philip Hone, mayor of New York, elected president of the company. The following day the respected financier John Bolton was elected treasurer. On March 17, 1825 the commissioners met and reported a net receipt of $74,207.59 from stock subscriptions.

It was with this modest working capital that the actual construction of the Delaware and Hudson Canal was begun.

CONSTRUCTION

July 13, 1825 dawned a bright sunny day. A great crowd of people from Ulster, Orange and Sullivan Counties had assembled, some of whom had driven twenty or thirty miles on jouncing, hard-riding buckboards on roads little more than trails. This exuberant crowd converged at a point midway between Ellenville and Wurtsboro near a little hamlet popularly called Summitville but officially known as Beatysburg.

The momentous event was the turning of the first shovelful of earth on the canal. This formal act by President Philip Hone initiated the letting of contracts for the early phases of construction. President Hone in his ensuing address set forth the objectives of the canal and the benefits expected to accrue after its completion.

Construction did not start simultaneously all along the line but only as the engineers completed their surveys and drawings. Actually the first contract was let a few weeks after President Hone turned the first shovelful of earth. It was for a seventeen mile lock-free section later to become known as the Summit Level.

On November 7, 1825, the exact route of the canal through the Rondout Valley for a distance of sixteen miles was settled, and on December 6, the last contract was signed for the line between Eddy's Factory (Eddyville), on the Rondout at tidewater, and Mongaup on the Delaware for a total distance of sixty miles. The sum of $30,000 was added to the estimate to bring the Neversink River into the feeder system via a three-quarters of a mile feeder ditch at the southwest end of the Summit Level, thereby affording an increased water supply. All operations were pushed with vigor until the heavy frosts of December 1825 compelled cessation of all work except deep cutting and rock blasting.

It is difficult today to form an adequate concept of the difficulties under which the rough and ready crews labored from 1825 to 1827. That they achieved such notable construction as they did is high testimony to their daring, originality and persistence.

It is well to remember that they had no means but hand drilling to make powder holes in the stone and nothing more than old fashioned black powder to do the blasting or *blowing* as it was usually called then. After the hole was laboriously drilled it was filled with black powder and a fuse inserted, tamped tight with clay considerably on the dry side. This so-called fuse was the culprit that kept the hiring boss recruiting new men for the job of *blower*.

The fuse was nothing more than a piece of twisted brown paper that had previously been saturated with saltpeter and dried. When all was in readiness the blower cautiously lit the fuse and raced for cover, hoping all the way that the charge would not sputter out. This saltpeter-soaked fuse was the tricky factor. Too much saltpeter in the brown paper and the fuse would burn too rapidly, preventing the safe escape of the blower; too little resulted in a delayed action. There were many men who vanished from the payroll when returning to relight a delayed fuse.

At the beginning one of the biggest contractors was Major Rensalaer Schuyler, who took the contracts for thirty locks. However, his work was not satisfactory to Wright, and he was relieved of his contract on November 16, 1825, receiving $1,500 for his losses and $1,200 for his labor and material. [5]

At the close of the year 1825, estimated work done aggregated $68,006. Even though the winter of 1825-26 was harsh, over 500 men were engaged in rock cutting and removal, and on February 28, 1826 contracts were let for sixty locks of hammered stone. As the blustery winter weather moderated into spring, over $123,000 had been expended, a good indication of the costliness of stone work.

By the late spring the activity upon the canal had reached such a point that the Kingston *Peoples' Advocate* stated that about twenty-five hundred men and two hundred teams were engaged on the Rondout and Summit sections. On the summit level, light or gravely soil sometimes caused the canal bed to leak water continuing as much as two or three years before stabilization occurred. In situations of this nature the channel was lined with a thick layer of stiff clay or *puddle*. A minor aggravation that developed from persistent leakage were law suits from canalside farmers whose fields would be mysteriously indundated overnight.

On a pleasant day in November 1826, a unique Masonic rite was performed at the consecration of the cornerstone of the Tidewater Lock in Eddyville.

The scene was witnessed by multitudes of workers and sightseers who lined both banks of the creek. In wagons and carts, on horseback and afoot, they had come from great distances to view the ceremony.

At Twaalfskill (Wilbur) a select company of officials and guests, many from the East and other points, boarded the *Morning Star* under the command of Captain Griffin, and with music, cheers and the booming of cannon, proceeded upstream to Eddyville where the Masonic rituals consecrating the cornerstone took place.

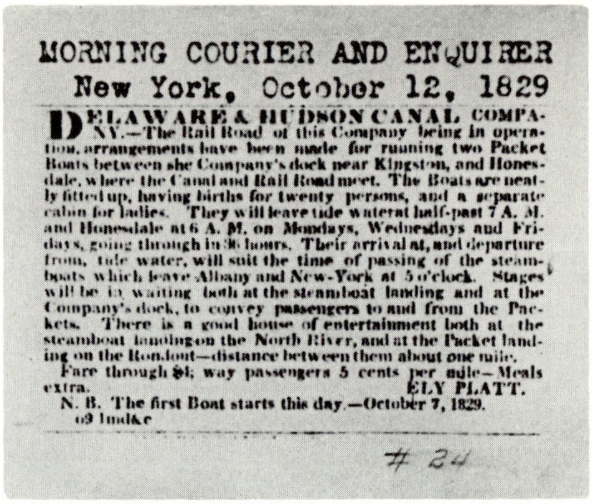

MORNING COURIER AND ENQUIRER
New York, October 12, 1829

DELAWARE & HUDSON CANAL COMPANY.—The Rail Road of this Company being in operation, arrangements have been made for running two Packet Boats between she Company's dock near Kingston, and Honesdale, where the Canal and Rail Road meet. The Boats are neatly fitted up, having births for twenty persons, and a separate cabin for ladies. They will leave tide water at half-past 7 A. M. and Honesdale at 6 A. M. on Mondays, Wednesdays and Fridays, going through in 36 hours. Their arrival at, and departure from, tide water, will suit the time of passing of the steamboats which leave Albany and New-York at 5 o'clock. Stages will be in waiting both at the steamboat landing and at the Company's dock, to convey passengers to and from the Packets. There is a good house of entertainment both at the steamboat landing on the North River, and at the Packet landing on the Rondout—distance between them about one mile.

Fare through $4; way passengers 5 cents per mile—Meals extra. ELY PLATT.

N. B. The first Boat starts this day.—October 7, 1829.

o9 lndkc

\# 24

Courtesy of PENNSYLVANIA HISTORICAL AND MUSEUM COMMISSION, HARRISBURG, PA.

Notice of inauguration, with schedule, of packet boat service on the D. and H. Canal between Kingston and Honesdale. First trip - October 7, 1829.

The official barge, followed by another brightly painted canalboat, continued on the high level for a few miles accompanied by the songs and cheers of the onlookers.

The animated spectators became so enthusiastic that they grasped the towline to relieve the gaily decorated horses. Speeches were read, the engineers and officials were cheered, and toasts were offered to the Delaware and Hudson Canal Company, to the workers, to Governor Clinton, and to American enterprise.

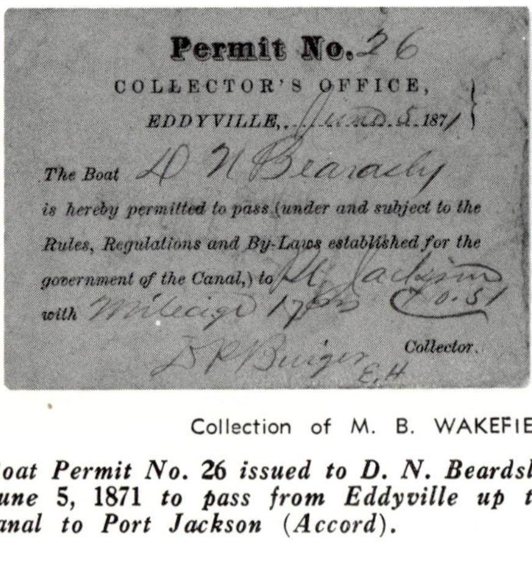

Permit No. 26

COLLECTOR'S OFFICE,

EDDYVILLE, June 5 1871

The Boat D N Beardsley *is hereby permitted to pass (under and subject to the Rules, Regulations and By-Laws established for the government of the Canal,) to* Pt Jackson *with* Mileage 17 $0.51

D R Burger E.H.

Collector.

Collection of M. B. WAKEFIELD

Boat Permit No. 26 issued to D. N. Beardsley June 5, 1871 to pass from Eddyville up the canal to Port Jackson (Accord).

Work on the Delaware section was urged forward in 1826, but, retarded by much wet weather, was deferred until the following season. On March 1, 1827, a notice signed by President Bolton was published, indicating that the building of thirteen locks on the Delaware River and twenty-two locks between the mouth of the Lackawaxen and the "Narrows" in Pennsylvania was open to contract.

The canal was completed largely by Irish laborers, and their working habits are an integral part of the canals' folklore. Whiskey cost a dollar for a two gallon jug, or sixpence a drink, and most workmen preferred it to water. Nothing made a man work harder for six days than the prospect of getting drunk on the seventh. [6]

Perhaps it was problems of this nature, whiskey and Irish mixed, that lay back of the release in the Monticello *Watchman* in August of 1827, stating that five hundred laborers were desperately needed on the D. and H. Canal in the Lackawaxen area. [7]

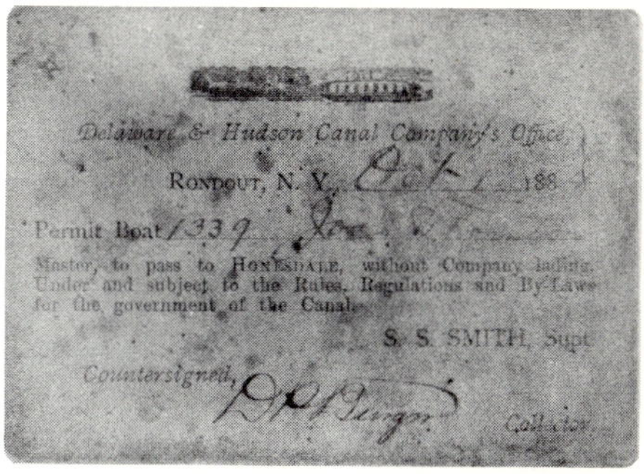
Delaware & Hudson Canal Company's Office

RONDOUT, N. Y. Oct 1 188

Permit Boat 1339

Master, to pass to HONESDALE, without Company lading. Under and subject to the Rules, Regulations and By-Laws for the government of the Canal.

S. S. SMITH, Supt

Countersigned,

Collector.

Collection of M. B. WAKEFIELD

Permit for boat No. 1339 issued to Jos. Thornton on October 1, 1884 to pass to Honesdale for a load of coal.

On April 1, 1828, the canal was considered functional by the managers. On April 22 this fact became apparent when the collector at Eddyville issued a tidewater certificate indicating the arrival of fifteen boats and one raft. Although the records indicate all the details of the flotilla, they conspicuously omit their point of origin and the cargo they carried.

During the summer all the minute finishing details were completed. By September a celebration committee was appointed including President Bolton, Mr. Hone, and Mr. Lord. The excitement became intense for the managers over the prospects of a ride down the entire length of the canal. A first trip was arranged, and on October 16, 1828 they departed Rondout for Honesdale on the *Orange Packet.* They were accorded a rousing reception when they pulled into the boat basin at Honesdale. This packet and the boats accompanying it were the first to navigate the canal for its entire length.

On December 5, 1828, a squadron of eleven boats from Honesdale arrived at Rondout. Each of the boats had on board ten tons of anthracite coal that was initially transported with considerable difficulty into Honesdale over a very rough and tortuous road through Rix's Gap, over the Moosic Mountains. This mountain range was the scene of feverish construction activities as the Gravity Railroad neared completion at the end of 1828. [8]

This first squadron was led by the boat *Superior* commanded by Captain Hickson and was followed by the *United States* under the command of Captain Kortright. The records fail to mention the names of most of the boats of the squadron and their captains. As this significant group of boats passed through the various towns and locks the festive occasion of the day demanded the drinking of toasts. With the frequency of locks in the Phillipsport area and at High Falls one wonders how the steadfast captains managed to keep their craft from altercations with the towpath, locks and berme banks as they proceeded down to tidewater. Providentially there were no Honesdale bound boats.

With a Kingston band on board the *Superior,* a tumultuous reception awaited the flotilla. The firing of musketry, the shooting of fireworks, and the blaring of many bands contributed to a din that was matched by the incessant tooting of the river craft. The somnolent north bank of the Rondout, called the Strand and later known as Bolton, would never be the same again.

By coincidence ten tons of this first coal was transferred to the sloop *Toleration* almost four years to the day after the *Tripler* had carried a load of coal to New York City for demonstration purposes. [9]

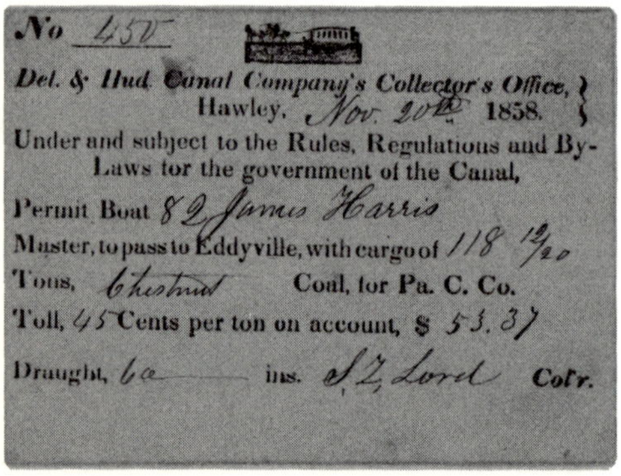

No 450

Del. & Hud. Canal Company's Collector's Office, Hawley, Nov. 20th 1858.

Under and subject to the Rules, Regulations and By-Laws for the government of the Canal,

Permit Boat 82 James Harris

Master, to pass to Eddyville, with cargo of 118 12/20

Tons, Chestnut Coal, for Pa. C. Co.

Toll, 45 Cents per ton on account, $ 53.37

Draught, 6a ins. J Z Lord Col'r.

Collection of M. B. WAKEFIELD

Permit No. 450 issued at Hawley in 1858 to the Pennsylvania Coal Company boat No. 82 for the passage of a load of chestnut coal.

The Lion and the Gravity

The Lion

On April 4, 1827, John B. Jervis, chief engineer for the Delaware and Hudson Canal Company, received a directive from the managers to survey and locate a railroad route from the proposed terminus of the canal to the mines at Carbondale.

In October a lengthy report was submitted to the managers recommending a double-rail railroad. The report also stated that there would be about eleven miles of almost level roadbed and for these sections locomotives of four or five tons could be used.

As a result of the recommendations, President Bolton, in his report of 1827, directed that resident engineer Horatio Allen [1] be sent to England to make purchases of railroad iron and locomotives. Upon arrival at Liverpool on February 15, 1828, Allen immediately struck up a deep and lasting friendship with George Stephenson, one of the pioneers of locomotive engineering.

By July 1828 contracts had been completed with Robert Stephenson and Company of New Castle for one locomotive, the *America,* and with Foster, Rastrick and Company for the *Stourbridge Lion,* the *Delaware,* and the *Hudson.* [2] The *America* was shipped from the Stephenson works on October 20, 1828, and arrived in New York on the ship *Columbia* on January 15, 1829, at a delivered cost of $3,663.30. It was set up on blocks at the foundry of Abeel and Dunscomb and allowed to collect dust until May 27 when it was demonstrated under steam. By May the *Lion* had also arrived and at this time the two locomotives were loaded aboard the steamboat *Congress.*

Originally two canalboats had been ordered for the Hudson River part of the journey. Engineer Jervis felt the risk of loss or damage would be too great so the *Congress* was hired for the locomotives and the eighty pairs of railroad car wheels. The cost of this operation was seventy-five dollars which included the towing back of the two canal boats.

At Rondout the *Lion* was transferred to a canalboat and on July 16, 1829, received a permit at the Tidewater Lock for the run to Honesdale. Records indicate the *America* never arrived in Honesdale; however, in 1923 a cylinder, at first thought to be from the *Stourbridge Lion,* was found at the Smithsonian Institution. A thorough check with the Science Branch of the British Museum identified it as a Stephenson product.

Further tracing revealed the cylinder had been deposited at the Smithsonian on October 15, 1890 by Lindsay and Early of Carbondale. Perhaps the *America* was consigned to limbo because of the attraction created by the glaring red lion painted on the front of the *Stourbridge Lion.* This artistic bit of elegance came about because the locomotive painter in England thought he detected in the rounded boiler head a farfetched resemblance to the king of beasts and so he painted it in.

The locomotive had a horizontal boiler, and two thirty-six inch stroke vertical cylinders with grasshopper type walking beams communicating motion to the iron-tired, oaken driving wheels.

People descended on Honesdale from all directions to see the first performance of the *Lion*. One other unscheduled and shocking performance however was in store for the assembled throng. When the starting cannon, tamped with too heavy a powder charge, was fired, it disintegrated in a spectacular blast of cast iron and bronze fragments. Also disintegrating was the arm of Alva Adams, the enthusiastic cannoneer who misjudged the amount of powder necessary.

With the roar of a cannon and cheers ringing in his ears Allen opened the throttle wide and began his historic trip. All along the route men waved their hats, small boys shouted, and women stared in amazement as he thundered past at the fantastic speed of ten miles an hour. Who would ever have believed that anything so large could move so fast without a horse to pull it?

The trial convinced Allen and the managers that the road was not suitable for locomotive power. The locomotive when delivered weighed in at seven tons, considerably more than the four or five tons originally planned on.

On September 9, the *Lion* was again tested in motion. This final test confirmed beyond a shadow of a doubt the engineer's earlier observations. In a report to President Bolton by Engineer Jervis it was pointed out that "The result has led us to the conclusion that our curved road with fifteen feet streaches (sic) will require additional support . . . The locomotive will I think answer our expectations, when we get the road firm enough to bear it. So far I think all the difficulties discovered can be easily remedied, though it must necessarily be at some loss of time." [3]

The *Lion* was run off the rails near the canal dock where it was left to stand, an object of "monster" proportions to the neighborhood kids who would make long detours to avoid going near it. With the advance of winter a rough wooden shed was built over it but nocturnal vandalism soon removed portions of it.

So it stood for fourteen years, in a semi-exposed condition until finally so many parts had been broken off and carried away that it was useless as a locomotive. The boiler was subsequently hauled away to the Carbondale shops of the D. and H. Company to be used in supplying steam for a stationary engine until worn out and consigned to the oblivion of the foundry yard of Lindsay and Early. Here the remains were eventually discovered and deposited with the Smithsonian Institution in Washington. A replica is on permanent display at Honesdale.

Because of the loss of time spelled out in Jervis's report to the managers and other speculative changes of an experimental nature, the company elected to proceed with horse power on the levels between the planes.

Across the trestle into the woods steams the Stourbridge Lion with Horatio Allen at the throttle. In later life Mr. Allen wrote his recollections of that memorable August 8th back in 1829.

Line Drawing by M. B. WAKEFIELD

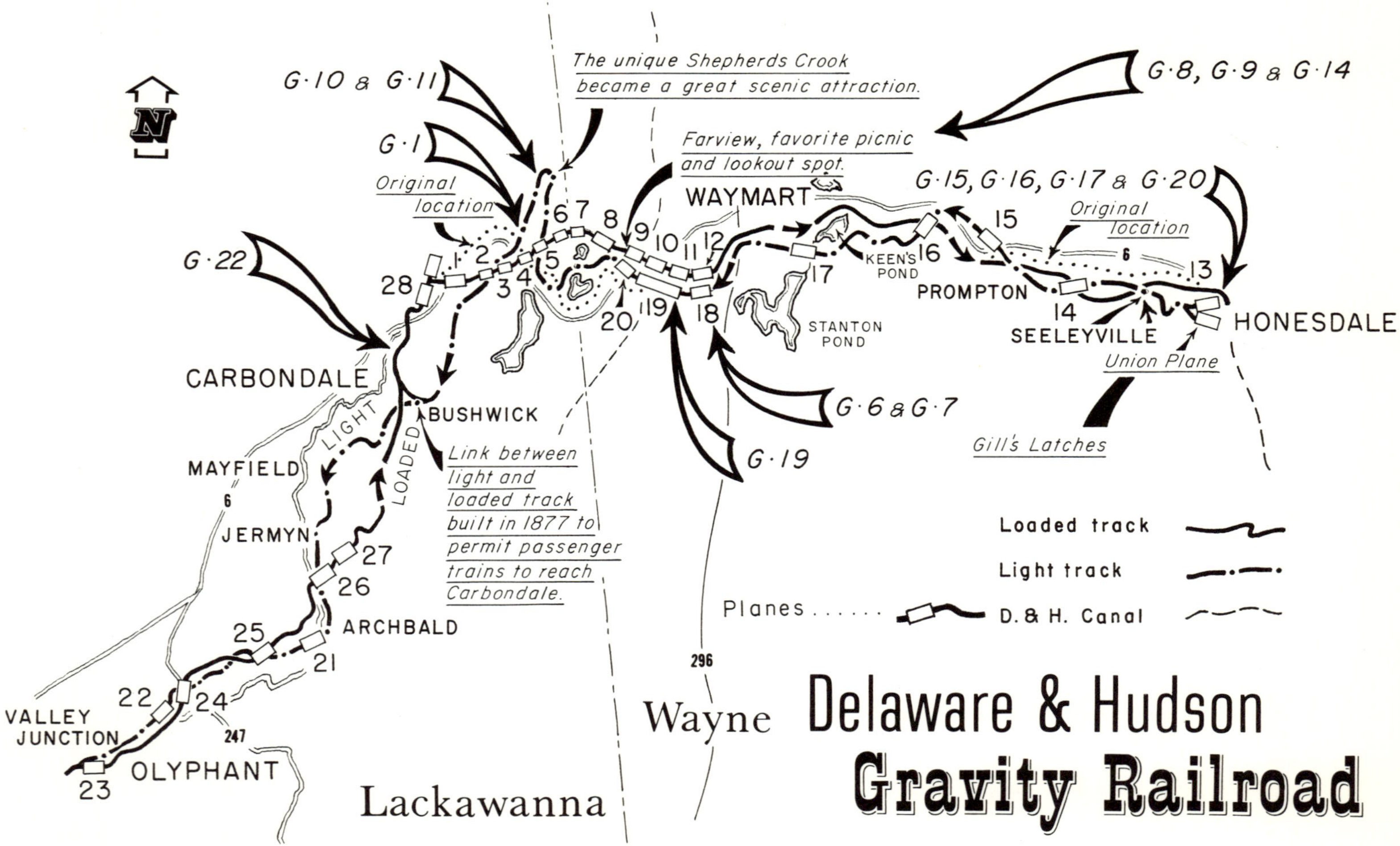

Delaware & Hudson
Gravity Railroad
G·10 & G·11
G·1
G·22
G·8, G·9 & G·14
G·15, G·16, G·17 & G·20
G·6 & G·7
G·19
The unique Shepherds Crook became a great scenic attraction.
Farview, favorite picnic and lookout spot.
Original location
Original location
Link between light and loaded track built in 1877 to permit passenger trains to reach Carbondale.
Union Plane
Gill's Latches
HONESDALE
SEELEYVILLE
PROMPTON
KEEN'S POND
STANTON POND
WAYMART
CARBONDALE
BUSHWICK
MAYFIELD
JERMYN
ARCHBALD
VALLEY JUNCTION
OLYPHANT
LIGHT
LOADED
Wayne
Lackawanna
296
247
6
Loaded track
Light track
Planes
D. & H. Canal

The Gravity

The work of locating the route for the Gravity Railroad was commenced in 1826 and by the summer of 1829 was completed from the mines at Carbondale to the canal basin at Honesdale, a distance of about seventeen miles.

The route from the mines required an ascent of nearly 950 feet in the first four miles to reach the summit of the Moosic Mountains. From the summit the descent to Honesdale was slighty longer due to a 20 foot elevation differential between the mines at Carbondale and the head of navigation.

The topography of the country through which it was constructed required great skill in engineering and much originality of plans. As 90% of the route lay through unbroken forests where timber could be very cheaply obtained, all heavy embankments for grades were dispensed with, and as far as practicable, without curves of too short a radius, heavy excavations were avoided. In an age of pick and shovel excavation work this decision had merit.

The Gravity Railroad development may be divided into three major phases: 1829 - The Original Gravity Railroad; 1841-1844 - Changes to the Gravity east of the mountains and changes of planes west of the mountains; and 1866 - Alterations of planes on the east slope providing independent movement of loaded and light cars.

Other changes took place from time to time, such as the 1856 reconstruction of the line from Carbondale to Waymart and various line extensions west of Carbondale. [4]

THE RAILS AND ROADBED

"The rails were placed in gains in the cross ties and secured by wooden keys or wedges in such a position that the space between the rails should be just the width of the gauge adopted, which was 4 ft. 3 in.

"The roadbed was not ballasted between the cross ties, and therefore there was nothing to hold the cross ties in position but their gravity, and the long wooden rails keyed into them; and nothing to keep the rails in place except the cross ties. Upon the top and at the inside edges of these rails, iron plates 2½ in. wide and ½ in. thick were laid to form the tracks for the wheels to turn upon and were made fast to the rails by large and long screws through holes in the plates made for that purpose. These holes were made longer lengthwise of the plates to permit their expansion and contraction by changes in temperature. After a little experience, the parts of the hemlock rails between the knots were found to be too soft for a firm bed for the iron plates, while the knotty parts remained firm, thus causing great unevenness in the surface of the track, to remedy which, strips of hardwood 1½ in. thick and 4 in. wide were spiked to the top of the wooden rails, and the iron plates were laid upon these strips. The use of screws for fastening the iron plates was also discontinued and iron spikes substituted." [5]

THE PLANES

"The inclined planes were constructed with a *single track*, but with turnouts, and a short 100 to 150 ft. of *double track* in the middle of each plane, to permit the cars moving in one direction to pass those moving in the opposite direction. These turnouts were provided with self-acting switches, or latches, so that whenever a car passed *out* of a turnout in either direction, the switch was left in the right position to turn the next cars moving in the opposite direction into the side of the turnout thus vacated.

DEL. & HUD. CANAL CO.

GRAVITY R. R.

Time Table & Connections.

TAKES EFFECT MONDAY, MAY 10TH, 1880.

Leave	HONESDALE,	6.40 A. M.	3.00 P. M.
	CARBONDALE,	8.20 "	4.50 "
	GREEN RIDGE,	9.00 "	6.00 "
Arrive	SCRANTON,	9.10 "	6.08 "
	WILKES-BARRE,	10.00 "	7.05 "
Leave	WILKES-BARRE,	—— A. M.	12.50 P. M.
	SCRANTON.	6.20 "	2.08 "
	GREEN RIDGE,	6.26 "	2.15 "
	CARBONDALE,	8.00 "	3.35 "
Arrive	HONESDALE,	9.30 "	4.50 "

R. MANVILLE, Supt.

Courtesy PENNSYLVANIA HISTORICAL AND MUSEUM COMM., HARRISBURG, PA.

R. Manville was appointed Superintendent of the Gravity on January 1, 1864, *replacing C. P. Wurts.*

"A Stationary steam engine was procured and used for motive power at each of the five long planes on the Western side of the Mountain. By means of a long chain extending the entire length of the plane, these engines drew up from three to five loaded cars, and at the same time let down a like number of empty cars attached to the other end of the chain.

"The three planes descending towards Honesdale were worked by gravitation, without the use of steam, by having the descending loaded cars draw up the ascending empty cars attached to the other end of the chain, and the velocity of their motion was controlled by the use of friction brakes upon the shaft of a large upright *Fan Wheel.*" [6]

After the disappointment with the *Lion,* it was determined that horses would be the power for the long levels descending into Honesdale. However even the use of the dependable old dobbin developed serious limitations as productivity increased at the mines.

". . . On the Summit Level one horse could not draw more than two loaded cars at a time. On the Six Mile Level (between Waymart and Prompton) the grade was such that loaded cars descended by gravity, and the cars were provided for a sufficient number of horses to ride with each descending train to draw the empty cars back, one horse thus being able to return four empty cars. These horses became so accustomed to riding *down* the grade that when by reason of ice on the rails, the cars required force to propel them, some of the horses clearly showed an unwillingness to go back upon the track and draw the cars in that direction.

"On the Four Mile Level (between Prompton and the Canal Basin), the grade was such that one horse could draw five loaded cars *down,* and the same number of empty cars back. The Four Mile and Six Mile levels had each a branch or side track for a short distance near the center; so that cars moving in one direction could pass those going in the opposite direction, and at these branches were the boarding houses for the *car runners.*" [7]

USE OF CHAINS

"A very few years' experience with the use of chains for conveying the loaded cars upon the planes convinced the manager of the Road that some material less liable to break must be substituted for the chains in use. Each train of loaded cars passing over the road was liable to be wrecked by the failure of any *one* of some 150,000 links composing the chains upon the several planes, and such accidents were so frequent that it became necessary to keep near the several planes portable Blacksmith's forges, which could, when neces-

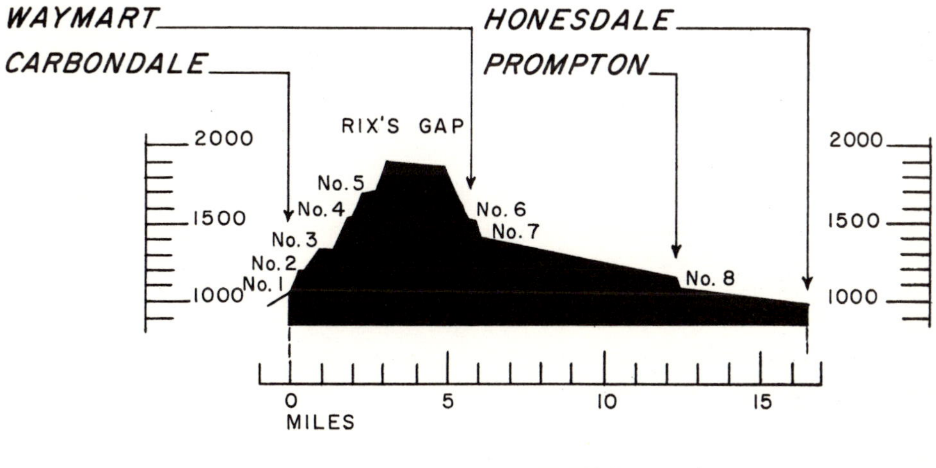

PROFILE
ORIGINAL GRAVITY RAILROAD
1829

G 1 Collection of JIM SHAUGHNESSY

Breaker at Plane #4. The original right-of-way of the Gravity cut across the upper part of Plane 4 about where the bridge is. The tall chimney at the top of the plane marks the location of the stationary engine house.

sary, be quickly taken to any position on the plane to enable the Smith to put in and weld a new link. To remedy this difficulty strong hemp ropes were substituted for the chains and thereafter such accidents were infrequent. After the invention of *wire rope* cables, these were substituted for the hemp ropes, and they are still in use on all the planes.

"This short experience also demonstrated that the use of horses to draw loaded cars up so heavy a grade as that extending from the mine to plane #1 was unnecessarily expensive. To remedy this, a plane operated by water power was constructed in 1837. This plane was 1400 ft. long, and, ascending from the mouth of the mine, nearly 80 ft. in height. From the head of this plane a new track 1500 ft. long, to the foot of Plane #1 was laid, on a descending grade of 1 in 250 ft. This improvement reduced the cost of transportation on that short section at least two cents a ton. Subsequent to these changes very little material alterations were made on the road before 1841." [8]

1841 - 1844 CHANGES TO THE GRAVITY EAST OF THE MOUNTAINS AND CHANGES OF PLANES WEST OF THE MOUNTAINS

By 1841 productivity of the mines had increased to nearly 200,000 tons per year, putting a considerable strain on the Gravity. All indications pointed to even greater production levels. The managers elected to make improvements to accommodate the projected increases.

"Mr. J. Archbald, Eng. in Charge, estimated that by a change of planes on the West side of Moosic Mountain and the long plane on the East side, so as to make them more nearly of equal length, and also the construction of a new road for the return of the empty cars from Honesdale to Waymart, together with a change in grade of the track for the loaded cars between Prompton and Honesdale, so that they would run by gravitation, the capacity could be increased 500,000 tons annually.

"This plan was approved by the Board and in the latter part of 1842, work upon these changes was commenced, and during 1844 they were completed. The track from the mine to the foot of Steam Plane #1, changed in 1837, . . . was not further changed." [9]

"For the return of empty cars from Honesdale to the Eastern foot of the Mountain at Waymart, another track was constructed entirely separate from the track used for loaded cars, with a series of inclined planes worked by stationary power, to draw the cars to such an elevation as would enable them to run by their own gravity from the head of one plane to the foot of the next. The unevenness of the country was such that in order to avoid having many high trestles it was deemed best to construct five such planes in that distance of 10 miles. These planes were so located that all except the one at the Basin could be arranged to use water instead of steam for power to work them." [10]

G 4 Courtesy PENNSYLVANIA HISTORICAL AND MUSEUM COMM., HARRISBURG, PA.

Locomotive C. P. Wurts, #2, was the second American locomotive used by the D. and H. in Pennsylvania. Built by W. Cook and Co., Scranton in 1860, she was retired from service in 1874.

"In 1848 there were indications that the high trestling on "Level #1"

G 2 Collection of M. B. WAKEFIELD

The "Lackawanna" #4, as originally built in 1862 at the Dickson Manufacturing shops in Scranton.

would not long remain safe for use, and it was decided to construct a new track up the valley of the Cadjaw Pond Brook and around the "Horseshoe" Bend, so as to relieve the necessity of maintaining the high trestling. This new track increased No. 1 Level by 2436 ft. in length, and although Plane No. 1 had 85 ft. added to its length and 15 ft. to its altitude, the grade of this new level was reduced to 47 ft. per mile. After the construction of this track around the Horseshoe, the high trestle was taken down. The use of water power to work these planes did not prove satisfactory and after a few years, stationary steam engines were erected to take the place of such water power." [11]

1856, NEW PLANE CONSTRUCTION - CARBONDALE TO WAYMART

It seems as though the Gravity was constantly involved in expansion or alteration projects. Only ten years after the major 1844 alterations another rebuilding program was underway.

Gerald M. Best relates that . . . "in 1856 it was necessary to rebuild the entire line from Carbondale to Waymart to increase the capacity of the road. A new series of planes and connecting levels between Carbondale and the summit was constructed adjacent to the old planes, and the number of planes increased from five to eight. Six of the new planes and levels were double-tracked. The other two planes were also double-tracked, but on this section, new tracks for light cars were constructed on descending grades to the foot of 'Return planes', by which they were elevated to the head of the next plane. These improvements

G 3 Courtesy of PENNSYLVANIA HISTORICAL AND MUSEUM COMM., HARRISBURG, PA.

D. and H. Canal Co. #4, the "Lackawanna" shown after rebuilding at the Carbondale Shops. She was in service until 1899.

were completed early in 1858, and at the same time, the strap rail previously used was replaced with iron "T" rail on the entire line, with wire rope replacing the hemp rope on the planes.

"During 1858 the line was extended from Archbald a distance of six miles down the valley to coal lines not previously worked. Fortunately, these lands, owned by the Company, were rich in coal of a very superior quality."

LOCOMOTIVES

In 1860 extension of the railroad was completed from Olyphant to Providence, a distance of four miles. This extension was unique in that it used locomotive power, the beginnings of present day locomotive power on the Delaware and Hudson Railroad.

Best lists five pioneer locomotives that were used on the 4 ft. 3 in. gauge railroad.

1. 0-4-0 "Major Sykes"	**W. Cook & Co. 1860**	**Collided with No. 11 in 1871 and wrecked. Rebuilt 1872 to 0-6-0., Scranton 1 - 1889**
2. 4-4-0 "C. P. Wurts"	**W. Cook & Co. 1860**	**Sold to Dickson Locomotive Works 1874.**
3. 0-4-0 "Honesdale"	**W. Cook & Co. 1861**	**Also named "Terrapin" and "Fine Plume". Ret. 1899.**
4. 0-6-0 "Lackawanna"	**Dickson No. 1 1862**	**Scrapped 1899.**
5. 4-4-0 "Isaac N. Seymour"	**New Jersey L. & M. Co. 1866**	**Exploded 3/10/79 at Carbondale and sold to Dickson Locomotive Works.**

G 5 Courtesy of PENNSYLVANIA HISTORICAL AND MUSEUM COMM., HARRISBURG, PA.

The D. and H. Canal Co. #5, "Isaac N. Seymour" built in 1866 by New Jersey Locomotive and Machine Co; Paterson, N. J. 4 ft. 3 in. Gauge. No. 5 was retired from service in 1879.

"Locomotive #4 was the first engine constructed at the Dickson Locomotive Works in Scranton and was the forerunner of hundreds of locomotives built at that works for the D. and H. Company. (G 2 and G 3) Locomotive #2, the *C. P. Wurts,* and later #5, the *Issac N. Seymour,* were used in passenger service which was inaugurated in the summer of 1860. (G 4 and G 5) Number 1, the *Major Sykes,* was rebuilt to an 0-6-0 standard gauge locomotive in 1872 at the Green Ridge Shop, and thereafter was used to haul Gravity passenger trains from Lookout Junction to Carbondale over a three-rail track. Two of the five locomotives remained on the Gravity lines until abandonment in 1899, but were used for switching purposes only at the terminals." [12]

1866 - PLANE ALTERATIONS ON EAST SLOPE OF MOUNTAINS

In 1866 authorization was provided to initiate construction leading to the separate operation of the loaded and light tracks on the east flank of the Moosic Mountains. This improvement would increase the capacity of the railroad 200,000 tons annually and would also require the construction of three new light track planes, namely Nos. 18, 19, and 20. The length of plane 18 was approximately the same as its companion Loaded Plane 12, but Plane 19 was equal to the combined length of Loaded Planes 10 and 11, while Plane 20 was approximately equal to Loaded Plane 9. G 6 and G 7

This new and most spectacular section was opened for traffic on April 21, 1868. These alterations, with the exception of minor changes, were the last made on the Gravity Railroad. From the inauguration of passenger service from Honesdale and Carbondale, this portion of the line quickly gained fame because of its novel construction and scenic beauty.

Farview, one of the points of highest elevation on this section, developed into a popular tourist and picnic spot. G 8. It was particularly popular with Sunday School groups for outings and annual events.

As Edwin B. Callaway relates, "The ride from Honesdale to Farview on the Moosic Mountain was free from smoke, cinders and dust. It was exhilarating. Residents of Honesdale looked forward to this trip with great joy and pleasure many weeks in advance of the date set for the Sunday School picnic.

"Trains would leave from the Delaware and Hudson passenger depot, Main Street, on special schedules. Open cars were sent to Honesdale from Carbondale to transport the picnickers to the grounds. Sometimes there were not enough open cars to take care of

G 7 Courtesy of J. VANCE HUNT

Switching activity at Waymart as seen from the highworks. The man standing by the string of empties gives some idea of their size and rugged construction.

Planes 18 and 12 at Waymart. The Waymart depot stands just to left of center.

G 6 Collection of M. B. WAKEFIELD

G 8 Collection of JIM SHAUGHNESSY

The substantially built depot at Farview, the summertime destination of many a picnic excursion over the Gravity. Planes 9 and 20 show in the foreground.

the people, and a car would have to be improvised. On one occasion a flat-bottom car, one used to convey lumber and other kinds of freight, was used. It was open all right, but the seats ran length-wise of the car and one of the wheels had a flat spot. With every revolution of the wheel there was a bump which kept the passengers good natured and furnished plenty of amusing conversation. G 15

"At Farview the picnic cars were run on a siding where they remained until time to make the return trip home.

"After a short steep climb over a red shale path the picnickers rushed to claim tables which were scattered around in shady places. Following the dinner (and one ate so much he could hardly walk), it was suggested that the group go to the first observatory, which was only a short distance away. Sauntering along the way some would step off the beaten path to search for wintergreen berries and low bush huckleberries. When the observatory was reached there were several steps to climb before reaching the top. G 9. Here was a platform approximately 15 feet square enclosed with heavy planks to prevent one from falling. The surface of these boards contained whittled initials of young men and women enclosed in a heart shaped etching . . . The visibility from this point on a clear day was far reaching, but did not compare with the upper observatory, a mile further up the Moosic Mountain from which point twelve lakes were visible, located in Wayne, Susquehanna and Lackawanna Counties."

The ride down to Carbondale was an exciting experience as the open cars followed the contours of the ravines and valleys, around the gentle slope of a hill, through a dense wood and out around the sweeping Shepherd's Crook. G 10. This engineering achievement was created as a result of a deep cut in the side of a hill with an extremely large fill that flowed down into the valley completing the wide loop. G 11

G 9 Collection of M. B. WAKEFIELD

After a session over the groaning board it was time for a hike up to the observatory for a view from the top of the Moosics.

This scenic piece of track was located west of Number 4 Reservoir, where it curved to the right and crossed over a loaded track near the head of No. 5 Plane, then continued for some distance. After a cut in the side of the mountain, an abrupt turn or loop was made on a huge embankment and the track continued in a southerly direction, recrossing the loaded track at the head of No. 2 Plane, to Carbondale. The loop was about 400 feet in diameter and 2000 feet in length and with a grade of 110 feet to the mile the track returned to the lower end of the loop to within 82 feet of itself, horizontally, 37 feet below its upper end.

G 11 Collection of G. M. BEST

Approaching the deep cut in the hill that preceeded the extensive fill of the Shepherd's Crook.

G 10 Collection of M. B. WAKEFIELD

View of famous Shepherd's Crook in the 1880's showing a cut of empties returning to the mines.

These alterations, with the exception of minor changes, were the last made on the Gravity Railroad. From the inauguration in 1877 of passenger service between Honesdale and Carbondale, to the closing of the line, the Shepherd's Crook - Farview ride was a popular scenic attraction.

G 12 Courtesy of PENNSYLVANIA HISTORICAL AND MUSEUM COMM., HARRISBURG, PA.

A string of the summer passenger cars taking their ease on a Gravity storage track.

PASSENGER OPERATIONS

A time card for the year 1896 indicated that during the summer months there were six passenger trains daily over the Moosic Mountains from Carbondale. The road used 36 excursion cars at the height of passenger operations. G 12 & G 13. Passenger operations also required that the line be operated throughout the winter months. At the summit level the tracks were subject to heavy drifting of snow which resulted in the erection of extensive snowsheds over the right-of-way. G 14

"Passenger trains were allowed speeds up to 35 mph, and a three man crew consisted of a conductor and two brakemen. The head brakeman rode the front platform, and for signalling purposes operated a foot-actuated bell located on the platform. This was used as a warning when approaching stations or crossings. When the head brakeman wished to retard the speed of the train, he applied his hand brake, and this would bunch the slack between the cars, indicating to the rear brakeman that he should apply his brakes. Presumably when the brakes were released, the head brakeman released his brakes and gave a bell signal, but no doubt the crews were so familiar with every foot of the road that it was second nature to apply the brakes at the right time and with the right force.

G 13 Collection of G. M. BEST

D. and H. Gravity Railroad Parlor Car 80 with its comfortable looking swivel chairs.

"Passenger cars in regular year round services were flat-roofed, without

G 15 Collection of JIM SHAUGHNESSY

Passenger laden excursion train about to start up Plane 13 ***out of Honesdale. The white horse to the right was used to pull the cars up to the foot of the plane. An overflow crowd has necessitated the inclusion of a specially equipped flat car.***

clerestories, and were equipped with two four-wheeled trucks. Baggage was carried in the head car which was usually a combination coach." 13

A reporter from the Port Jervis *Tri-States Union* describes a typical trip over the Gravity in the October 7, 1879 issue.

"From the Delaware and Hudson depot at Honesdale, one pleasant afternoon we stepped aboard the train, which was composed of three finely appointed cars, one baggage and smoking, one passenger coach, and one open car, similar to those used on the Brighton Beach Marine Railway. G 15

"Very soon after three o'clock the train glided out from the covered depot, and after going a few yards stopped at the foot of the first plane. G 15 & G 16. We were anxious to know how the train

G 14 Collection of GERALD M. BEST

A portion of the extensive snowsheds near Farview, the summit of the Moosics.

would ascend the hill, which stretches at an angle of sixty degrees for one quarter of a mile above us. Soon an employee pulled three times on a rope, which runs up the hill, beside the track, to the engine room, and rings a bell.

"The engineer started his stationary engine slowly, and the strong endless cables, which run over pulleys between the tracks began slowly to move. A train hand stood at the front platform with a *sling,* which had a large hook at each end, in his hand. A chain of large links appeared in the endless cable and he inserted one of the hooks in a link and the other in a link in the front platform of the car, and we were off. Up, up the hill we ran at the rate of fifteen miles per hour. There is no visible motive power, but the trains are drawn up the hill by means of the endless cable, which winds and unwinds about an immense iron drum in the engine room.

G 16 Collection of JIM SHAUGHNESSY

Plane* 13 *with a cut of three empties starting up the incline past one of the many storage piles of coal. Note the elevated trestle-like structure on which the coal cars are rolled for dumping.

G 17 Courtesy of PENNSYLVANIA HISTORICAL AND MUSEUM COMM., HARRISBURG, PA.

The stationary engine at the head of Plane* 13 *used for pulling cars up from Honesdale.

"After gaining the hilltop, the cable slacks and by a dexterous twist the brakeman disengages the sling from the cable. G 17 and G 18

"We didn't have time to long admire the scenery, for the train was soon underway, controlled only by brakes. Rapid speed was quickly attained and we were rushing around the "horseshoe". Seelyville could be seen in the distance, and after a run of four miles down a grade of fifty feet to the mile, we approached Prompton, which is a small hamlet where the Wayne County Normal School is situated. The plane at this point was quickly ascended in a manner

G 18 Collection of JIM SHAUGHNESSY

The stationary engine house at the head of Plane "F". All of the planes were numbered but a few were also known locally by letters such as Planes "G" and "F" between Carbondale and Olyphant. Henry Soete of Honesdale recalls hearing certain planes being referred to by letter in the Olyphant area.

similar to that already described and we were on the hill tops again, breathing in from our seats in the open car the beauty of the Wayne County hills, beautiful in their autumnal tints. Here we again glided down the road, around sharp curves, through bits of wood, through meadows and finally reached Keene's Pond - - a once well-known spot for pickerel fishing, and into whose waters the fisherman used to delight to throw his line. Now it presents the appearance of a mud pond, and stumps and brush predominate.

G 19 Collection of M. B. WAKEFIELD

Planes No. 19 and 11, shown here, were the longest on the Gravity.

"Plane number 19 we found the longest on the road. G 19 . . . Stanton's Pond, a resort for picnic parties, could be seen over beyond the hill at the left

Early view of the upper boat basin in Honesdale. At the upper end just past the complex of trestling is the feeder point from the West Branch of the Dyberry. The slightly inclined track to the left on the hill leads to the storage pile area used in the wintertime. To the extreme right is located the Canal Co. office. The denuded hillside dates this as an early view of Honesdale. Later views indicate considerable foliage.

G 20

Collection of JIM SHAUGHNESSY

of the plane. At Waymart, a village of some importance, we reached the foot of the Moosic range of mountains. Three planes were ascending in rapid succession, with but a few yards intervening between the top of the one and the foot of the other. We were finally at the summit of the Moosic Mountains - - two thousand feet above the sea.

"Eight planes had been ascended and we were ten miles from Honesdale. The road ran gradually from the top of the mountain without encountering any planes into Carbondale, six miles from the summit. It is the first mining town in the Lackawanna Valley beyond Honesdale.

"After waiting twenty minutes at the Summit, spent in viewing the manner of running loaded cars down the planes, the return train was sighted. We boarded it and the train commenced its descent of the mountain into Waymart. The manner of running trains down the planes is nearly as novel as that of their going up. The "sling" is attached to the rear car, and the cable and the train runs down the incline by its own gravity, but its speed is regulated by a large fan, whose atmospheric resistance is very great, situated at the head of the train. At Waymart a clear stretch of ten miles of road was reached, which ends at Honesdale. The brakes were loosened, the train gradually gained momentum, and soon we were making the curves and bounding along at thirty miles per hour . . . at five o'clock we reached Honesdale, and it was with a feeling of regret that we left the cars." G 20

COAL OPERATIONS

During the 1890's the tonnage of coal over the Gravity continued its steady decline from the all-time high of 4,097,218 tons reached in 1883. Management was still optimistic, however, and continued the "business as usual" attitude even though trains were getting noticeably shorter.

When the trains were running empty westbound, between Farview and Archbald, they would be assembled into 75 car trains with a maximum speed of 20 miles per hour. Special safety features were built into the planes to protect life and limb against runaway trains. These devices were a type of spring latch and were spaced at intervals along the planes. They were open derails in their normal position, and closed only to permit the passage of the upbound wheel flanges. G 1. Occasionally when a cut of cars did break away, the coal would be liberally distributed about the plane and the cars reduced to junk.

During the last 25 years of operation the company saw fit to gradually replace the small coal gondolas with cars having a capacity of five tons. The cars were all of the four wheel type with link and pin coupling. At the height of operations the Gravity inventory listed 5000 such cars in service. G 21

The small movement of anthracite over the canal and Gravity in 1897 forcast its approaching abandonment. During 1898 management decided to replace the Gravity with standard gauge track for direct connections with the Erie R. R. in Honesdale. By standard gauging the Gravity the company could then run coal cars of 20 tons or more capacity directly from the breakers to the consumer.

On January 3, 1899, the railroad was operated for the last time in its entirety as a gravity line, and then only for the purposes of running off anthracite that had accumulated on its tracks and for bringing cars to convenient places for disposal. G 22. On the following day the fires in the boilers of the stationary engines were drawn on portions of the line and its operations as a locomotive road commenced over the loaded track between Honesdale and Waymart.

The change from the 4 ft. 3 in. gravity gauge to that of 4 ft. 8½ in. was effected by moving each rail out one-half of 5½ inches. For a year prior to the third of January, men had been busy lowering the track to increase the overhead clearance at the breakers at Jermyn, White Oak and elsewhere along the line since the standard gauge cars at that time required at least ten feet head room. G 23

G 21 Collection of M. B. WAKEFIELD

A long string of empties sidetracked in an immobile position by a derailed front truck and a piece of chain. Possibly these cars were stored in this manner shortly after abandonment of the Gravity in preparation for scrapping operations.

G 22 Courtesy of PENNSYLVANIA HISTORICAL AND MUSEUM COMM., HARRISBURG, PA.

All official sources record January 3, 1899 as the date when the last trains operated over the entire Gravity. However a scrawled notation on the back of this old photograph notes the last passenger train out of Carbondale to Honesdale as leaving on January 21, 1899. Edward Hubbard was conductor on that last run.

The summer excursion cars as well as the regular passenger cars were bought up by people along the line for use as out-buildings or summer cottages.

The Delaware and Hudson Company records indicate that 4012 coal and other cars were scrapped in 1899 and 608 cars in 1900. The steam railroad continued to operate until 1931 at which time coal traffic was just a trickle and passenger traffic was virtually non-existent.

On August 20, 1931 the last standard gauge train made its melancholy run over the original Gravity right-of-way, almost 102 years to the day since the opening of that pioneer railroad.

G 23 Line Drawing by M. B. WAKEFIELD

In this post-canal era scene the bridge has been strengthened by pilings driven down into the abandoned canal bed. Along the left hand edge you will note that the tracks leading to Plane 13 have been cut by a newly laid standard gauge right-of-way.

The Canal, its Locks and Boats

The original Delaware and Hudson Canal was built to contain water to a depth of four feet with a width of 32 feet at the water line. The berme bank and tow path sloped down to a 20 foot width on the bottom. [1]

As the productivity of the mines in and around Carbondale increased by great strides, considerable demand was put upon the canal, its locks and boats. The first consideration by the managers of the desirability of enlarging the canal occurred on March 24, 1841. Increasing competition was a prime factor in motivating the change so that a greater volume could be moved to market at lower cost resulting in a more competitive price in the market place.

The word went out from the managers in September 1842 to initiate the enlargement of the canal. The canal proper would be deepened to a depth of five feet, thereby permitting boats of 40 tons capacity. This would be achieved by raising the berme and towpath sides as well as deepening the bottom. Because of the need to keep the boats moving, work was not commenced until the end of the boating season in November 1842 and continued through the winter until the opening of the 1843 boating season. It wasn't until the season of 1844 that the job was completed, however the full five foot head of water could not be permitted immediately.

The disruption of the stability of the canal bottom, the breaking of the angle of repose of the upper berme banks and the settling of the new earthen fill demanded that the additional foot of water be eased in gradually. As the new alterations settled and full operations were underway, the managers began to realize vast savings. These physical improvements immediately started the managers thinking about further improvements, with the result that 1845 saw still another enlargement program started.

The goal was to increase the water level to five and one-half feet, permitting the use of boats of 50 tons capacity. These constant enlargements, seemingly one right after the other, no doubt made the boat builders at Wilbur, Ellenville, Phillipsport, Hawley, and other points exceedingly happy. About 100 boats of 50 tons capacity were built for the anticipated enlargement, long before they could be adequately absorbed into normal canal operations.

The year 1847 saw the completion of the latest enlargement project. Washouts and an extended summer drought during the two year enlargement period caused the managers great concern. It also hindered the use of the new 50 ton boats to their fullest capacity. But for all the problems encountered, the annual report for 1847 noted that the enlargements had cost slightly over $232,000 and that savings were totaling out to $300,000. Most important of all, however, was the capacity of the canal, which was raised to 500,000 tons.

On November 17, 1847, President Wurts recommended further enlargement of the canal. This project estimated at $1,105,000 would permit boats of 130 tons capacity and would involve not only increasing the boating depth to a full six feet but would necessitate widening it. This final determination to widen the canal would involve substantial changes to the locks and aqueducts. It was urged by the managers that the work be commenced in 1848 with the hope that the entire job could be substantially completed by the spring of 1850.

Another distinct advantage was that with boats of such increased capacity and seaworthiness it would now be possible to ship direct to Hudson River destinations and New York City without the necessity of the time-consuming and costly coal transfer at Rondout. It was estimated that by this enlargement the freight charges on coal would be reduced to fifty cents a ton.

The enlargement was started in 1848 and continued into 1849. By the end of 1849 the changes were completed with the exception of some additional work to be done on the trunk of the canal in order to provide the full six feet of water. When this ultimate depth was obtained, the annual capacity of the canal became five times that of the year 1842.

During this last enlargement the sloping sides of the canal were extensively cut back and laid up with dry stone walls. These walls are still visible today (1965), particularly those along the towpath side. Many have collapsed due to their almost vertical construction. CLB 1

As mentioned previously, the locks were also enlarged, along with the construction of four suspension aqueducts. This resulted in two canal alignment changes, one at the confluence of the Lackawaxen and Delaware Rivers and the other, a rather extensive re-routing at High Falls, in Ulster County.

The locks on the Delaware and Hudson Canal were originally built to a width of nine feet six inches and a length of 75 feet. Built to accommodate boats of 20 tons capacity, the locks were none too wide for the eight foot five inch width of the early boats. It was truly said that the boats slid into the locks like a hand into a glove.

These pioneer locks, the stonework of one of which still stands just north of Alligerville, were of the balance beam type. R 10, R 11 and R 12. This type of locking required that each gate be operated independently by hand. Needless to say this system was very time-consuming and expensive in that two men had to be employed for the operation of the lock.

CLB 1 M. B. WAKEFIELD

A segment of the stonework lining that was erected along both sides of the canal during the enlargement program of 1848 - 49. This berme bank section is on the upstream side of the Beaver Brook Aqueduct along the Delaware River.

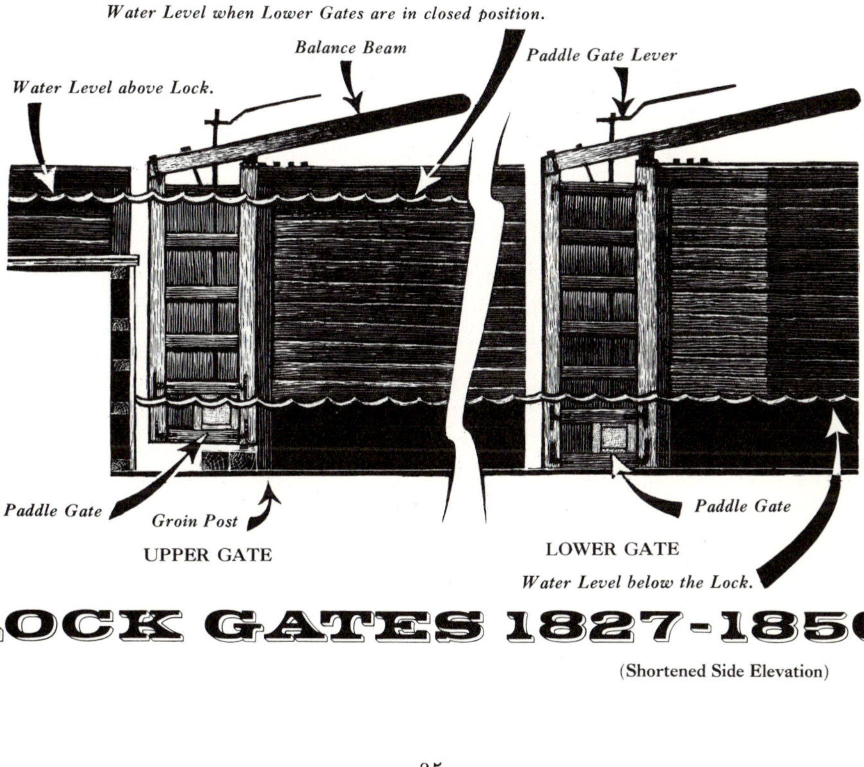

LOCK GATES 1827-1850

(Shortened Side Elevation)

Some locks on the canal were constructed of hammered stone. A few fine examples (circa 1856) still stand, affording the traveler an opportunity to inspect at close hand the precise cut and fit of the stoneworkers' art. R 13

Other locks were wood lined with a backing of stone not of the cut-and-fit variety. Virtually all of the wood lining has disappeared, having been torn out by area farmers or burned just as it stood. Careful inspection of some of the locks will reveal the charred stumps of the old planking. L 28 and L 32. By actual count there were 13 locks of the stone and masonry type and 95 of the composite or wood lined type. The total number of locks was 108. Some sources list 107, others as many as 110 (possibly counting the two weigh locks), but there is hardly an old *canawller* whom the author interviewed who didn't sum up with, "Yep, one hundred and eight locks and one hundred and eight miles."

The first boat floated on the canal was built and owned by Colonel Peter Cuddeback of Huguenot. It had been transported from Newburgh in sections on wagons, put together and launched at Cuddebackville. [2] One of the first big

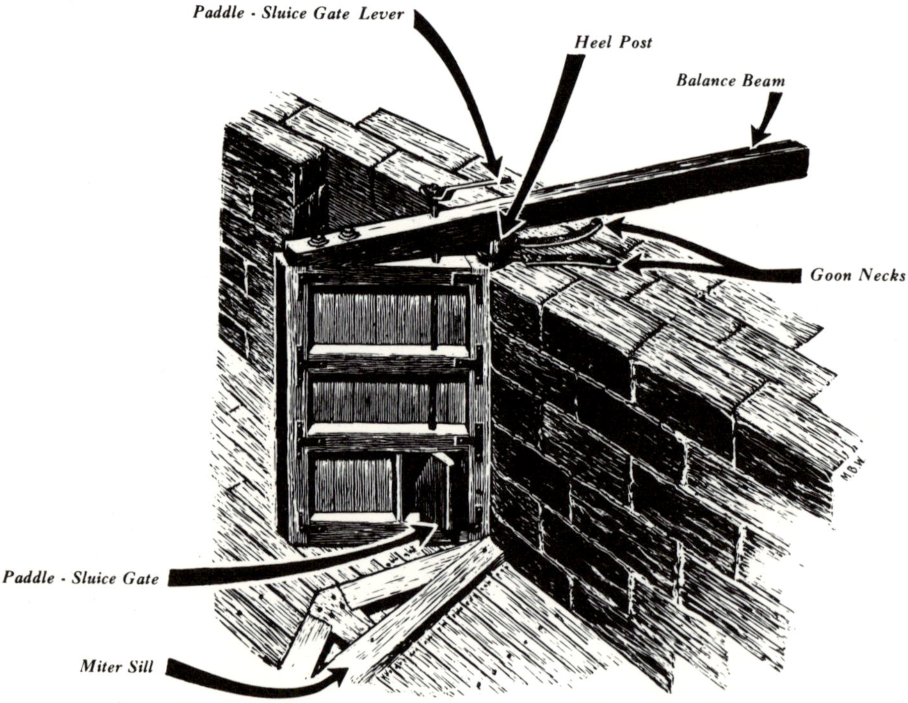

CANAL LOCK GATE
1827-1850

CLB 3 Courtesy of ELLENVILLE PUBLIC LIBRARY

Boats were launched sideways into the canal as shown here at Ellenville. Launchings were appreciated by the public as witness this crowd complete with a Civil War Veterans' contingent. Lock #31 shows in the background.

boatbuilding contracts on the canal was let at Rondout on October 23, 1833. This order for 100 canalboats for delivery the following spring added impetus to the growing boatbuilding industry at Rondout. All winter long the hills rising up from the docks and boatyards echoed to the sound of the adz and caulking hammers. CLB 2. These first boats, of 20-ton capacity, were later called *flickers* and ruled the waves of the D. and H. Canal until 1844 when heavier 40-ton capacity boats began to take over. CLB 3

During the transition period between the flickers and the 40 tonners, a considerable amount of antagonism developed between the captains of the two boat types. The newer boats, with slightly blunted bows, were forced by virtue of their design and greater load capacity to proceed at a slower pace. They also had a tendency to create a larger wash which was damaging to the raw earth exposed as a result of the enlargement project. The lighter and faster flicker could not pass the heavier boat when overtaken unless granted permission by the captains. Many flicker operators who could not afford the larger

CLB 2 Collection of M. B. WAKEFIELD

Typical boat building scene near Creek Locks on the Rondout section.

CLB 4 Courtesy of ELLENVILLE PUBLIC LIBRARY

A loaded boat with the bunker area in front of the cabin "hipped" in order to increase coal capacity. Note typical canal wagon bridge in background.

40-ton boats *hipped* their boats to increase the capacity. Hipping was achieved simply by raising the sides of the coal bunkers to permit a larger load of coal. CLB 4

By the year 1847 the flicker had pretty much disappeared from the canal. The heavier 40-ton boat, which gradually took over as stabilization of the canal banks continued, cost the canal company between $350 and $375 each. These 40-tonners would then be sold to the canal boatmen for $400. As the size of boats grew with the further enlargement of the canal, so did the cost of a boat until, near the end of operations, a new boat, right off the ways, would cost the canaller about $1800. The company would furnish him with the money if he didn't have it and would then deduct fifteen dollars from each trip to pay it off.

By 1850 the enlargement of the canal had progressed to such a point that boats of 98 tons capacity were the rule. In 1852 the locks were enlarged to 90 feet in length and a width of 15 feet. Some locks had a lift of 8 feet and others as much as 12 feet but the average was 10 feet.

At the same time that the locks were being rebuilt, still another canal enlargement was being completed that had been authorized by the managers in 1850. This most recent of the series of widening and deepening efforts would permit boats of 130 tons and was expected to increase the tonnage of the canal to a possible 1,000,000 tons. This confidence of increased tonnage down the canal was borne out by a jump to 425,993 tons in 1853 and the ultimate attaining of a million-ton goal in both the 1855 and 1856 boating seasons.

During the period of canal and lock enlargement, major relocation projects were undertaken at High Falls and Lackawaxen. Both changes evolved as a direct result of the building of three suspension bridge aqueducts by John A. Roebling (see Delaware and Lackawaxen Aqueduct Chapter). The High Falls realignment took the canal directly up the hill via six locks, No's 15 to 20 inclusive, each having slightly over a 12 foot lift. This project resulted in one less lock and necessitated the renumbering of all the locks from this point to the Delaware Aqueduct.

About 1865 another modification was made to the locking system. The upper gate was replaced by a drop gate and at the same time hand operated machinery was installed over the lower gate which controlled both the upper and lower gates. This new concept eliminated one man, resulting in an operational saving as well as speeding up the locking through of boats. CLB 5

CLB 5 M. B. WAKEFIELD

The lower gate and gate house at Lock #44 north of Phillipsport from a scale model by the author. The ratchet arms on the gates open and close the paddle gates under water. Gates are opened by a cog-and-wheel mechanism in the gate shed.

CLB 6 Collection of MALCOLM BOOTH

Light boat riding out of Neversink Lock #51 under the lock tender's bridge. The size and bulk of the boat clearly indicates the close clearances within the lock.

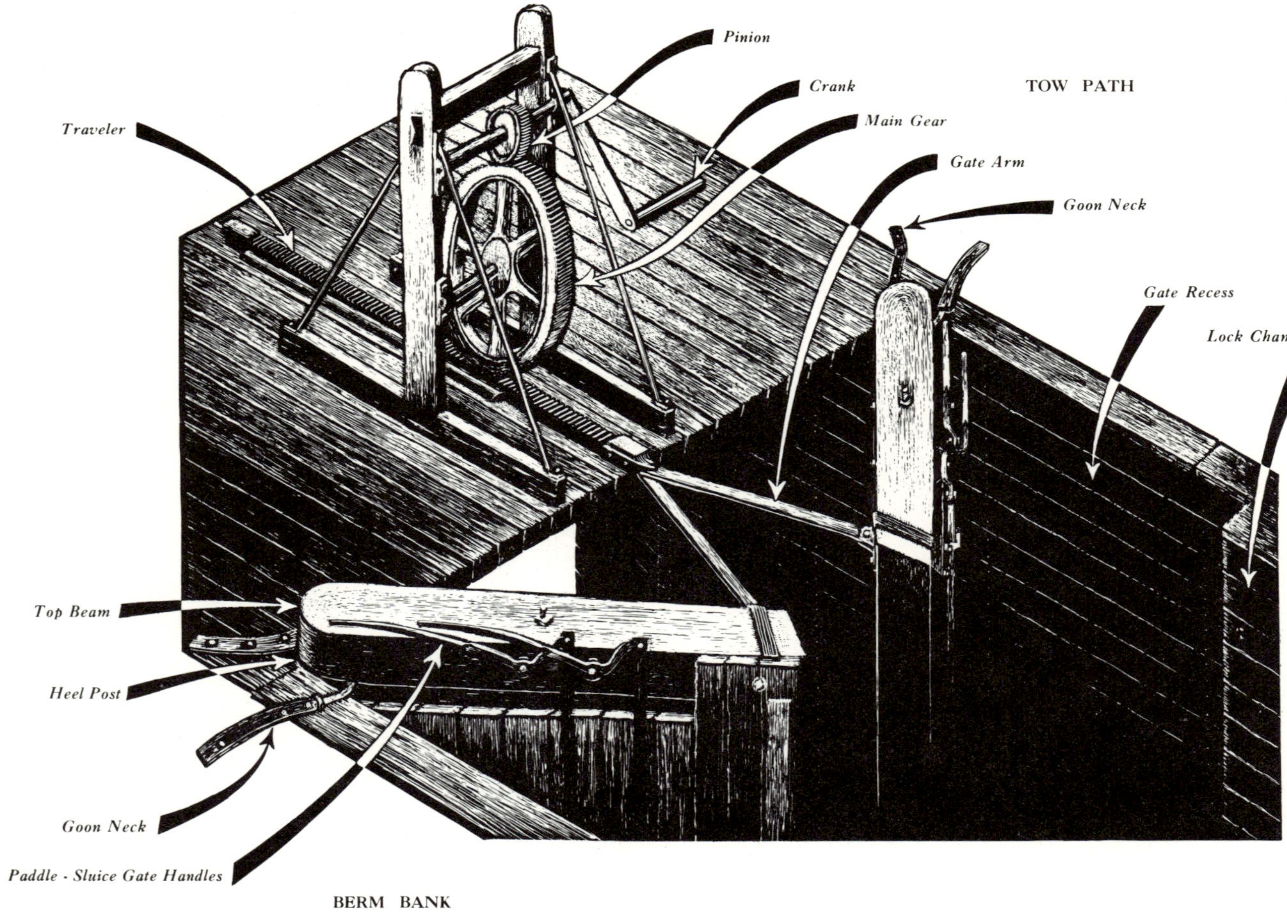

CANAL LOCK GATE 1850-1899

In locking down the canal, the loaded boat would ride over the drop gate, which was lying horizontal on the bottom of the canal, to a position for snubbing.

A.

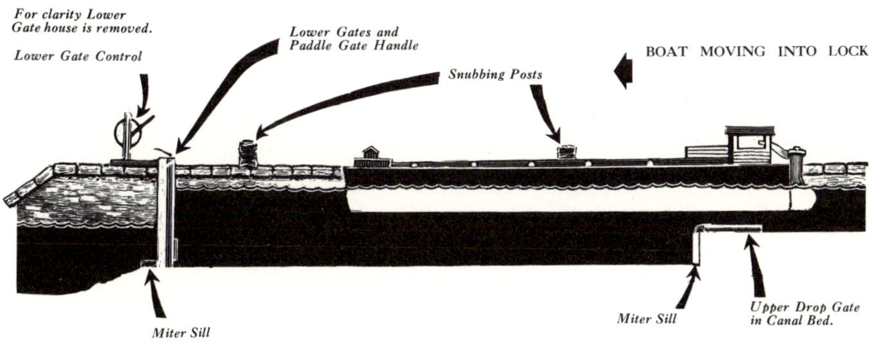

When the boat was snubbed securely, the lock tender, from the bridge over the lower gates, would gradually open the paddle gates in the lower part of the gate below water level, and at the same time wind up the drop gate using rods and chains located in the bottom of the lock.

B.

Generally the current created by the "cracking" of the paddles on the lower gates would be sufficient to draw the drop gate securely into a vertical position. The winch could now be tightened and the ratchet secured on the cog.

C.

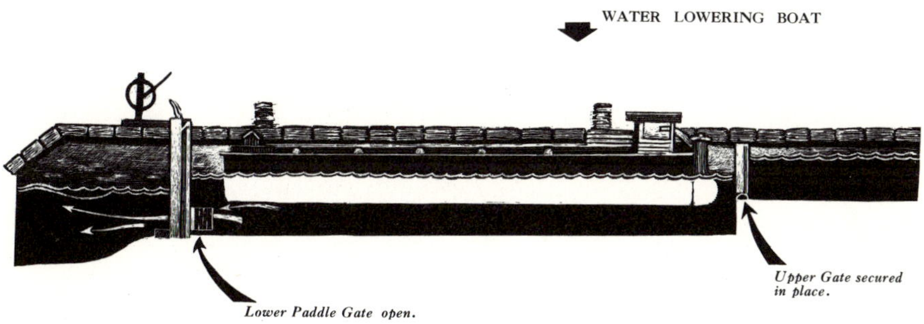

The paddles in the lower gate could now be opened fully allowing the water in the lock to flow out into the lower canal level. When the water in the lock reached the level of the water below the lock, the lower gates were opened and the boat glided out under the lock tender's bridge.

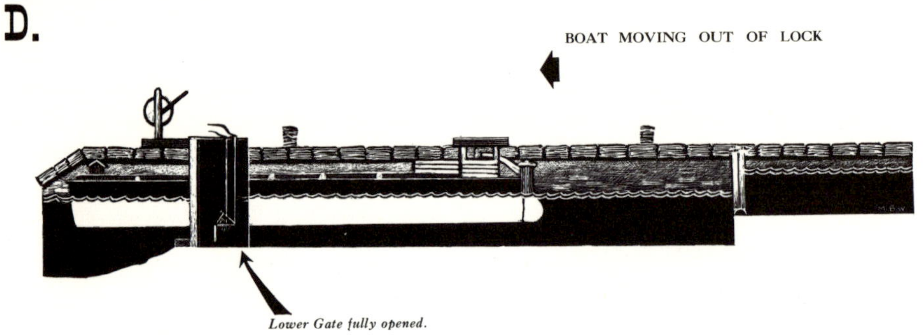

If locking through just after a rainy spell the lock tender could briefly open the paddle gates on the upper drop lock resulting in a swell of water as the boat moved out, thereby achieving the top speed of three miles per hour in a short distance. CLB 6

In locking light boats up, the lower gates would be closed after snubbing and the paddles on the drop gate opened gradually. When the water in the lock reached the level of the water above the lock, the drop gate was permitted to fall back into the bed of the canal.

The late DuBois Weber described the locking through of a light boat. "The horses would pull up past the lock until the tow rope flew off the timberhead as the boat moved in under the lock bridge over the lower gates. I would then pull out the tiller and throw up the light boat lever line to the towpath boy. He would then take this line and put it a couple of turns around the snubbing post continuing to slacken off the line while the boat moved into the lock. When the boat was far enough in the lock, the boat stopped and the gates closed behind. The paddles were opened in the upper gate and when the lock was filled the towline was put on the little timberhead (bow) again and we started off down the canal."

CLB 7 M. B. WAKEFIELD

A typical lock tender's house at Lock #35 on the Ellenville section. In this instance the building also served as a canal store.

The lock tender generally had a small house adjacent to the lock for which he was responsible. This house and a small plot of land was owned by the canal company and sometimes the lock tender would operate a store nearby or run a farm. CLB 7 and N 1. The primary responsibility of the lock tender, aside from the efficient locking through of boats, was maintaining the water level of the canal below his lock, the water level above the lock being the responsibility of the lock tender at the preceding lock.

There were two weigh locks on the D. and H. Canal, one at Hawley and the other at Eddyville. L 19 and R 33a. When moving through the Tidewater Lock in Eddyville at the beginning of the boating season, the light boat was moved into the weigh lock and the water removed allowing the boat to settle on large scales in the bottom of the lock chamber, thereby recording the basic unladen weight of the boat. The boat's name, number, and weight was listed at the collector's office, the water readmitted to the lock and the boat floated out. When the boat returned loaded to Eddyville, it was again weighed and the difference noted, thereby indicating the tonnage of the cargo.

As Mr. Weber described the procedure when arriving at Eddyville, "You would have to stop just above the weigh lock and the weighmaster would come aboard and run a stick down the pumpbox holes to measure the water in the bilge. If there was more than half an inch of water you had to stay above the weigh lock until you got the water pumped down to within a half an inch; then you could proceed into the weigh lock."

Boat development grew until they could carry 125 to 145 tons of coal. The first new boat concept to come along was the *futtock* boat which was designed

CLB 8 Collection of JAMES A. O'BOYLE

A midsummer's day scene of a "squeezer" boat as viewed from a canal bridge. Note the dead air space in the middle of the boat. Evidence of the umbrellas and the striped bunting would suggest this as an excursion outing.

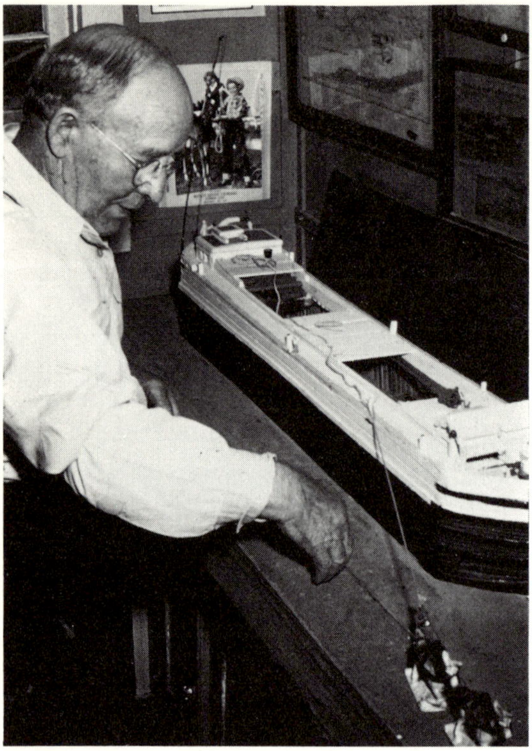

CLB 9 M. B. WAKEFIELD

The late DuBois Weber and the scale model canal boat he constructed out of scrap wood. Mr. Weber had the model peopled with scale model figures representing each member of his family.

to carry 125 tons. Another boat idea that emerged was the sectional boat or *squeezer*. CLB 8. This was a blunt-nosed monster which was divided in the middle and came apart in two sections that could carry a total of 130 tons. This facilitated two separate shipments in one boat for different destinations on the Hudson River, or the carrying of two different types of coal. At Rondout the boats would be disconnected and sent their separate ways. Although their capacity was larger they did not gain overwhelming popularity with the canalmen. Eight inches of dead water in the center between the boats resulted in a drag that necessitated a greater length of time in reaching tidewater. The drag on the boats also resulted in extra feed for the horses. Meanwhile the lighter boats were completing more trips before the end of the boating season. Accoring to DuBois Weber the last squeezer boat to run on the canal was operated by a woman by the name of Miss Gordon. CLB 9

The Port Jervis *Tri-States Union* of September 1879 reported the popular acceptance of the *log bilge* boat which originally came on the D. and H. scene in 1859. This style carried about the same amount of freight as the squeezers, and as they towed more easily were much preferred by both the company and the canallers.

From the time of the final enlargement to the abandonment of the canal, forty-nine years later, the canalboats all met the same specifications obviously dictated by the size of the locks, which were 90 feet long, 14 feet 2 inches wide at the bottom, and 14 feet 4 inches wide at the top. [3]

The most important factor in the development of all these larger capacity boats was their ability to carry a greater load with a shallower or no greater draft of water.

The ease of living on board a canalboat depended in large part on how well organized and compact the essentials were arranged. The average cabin

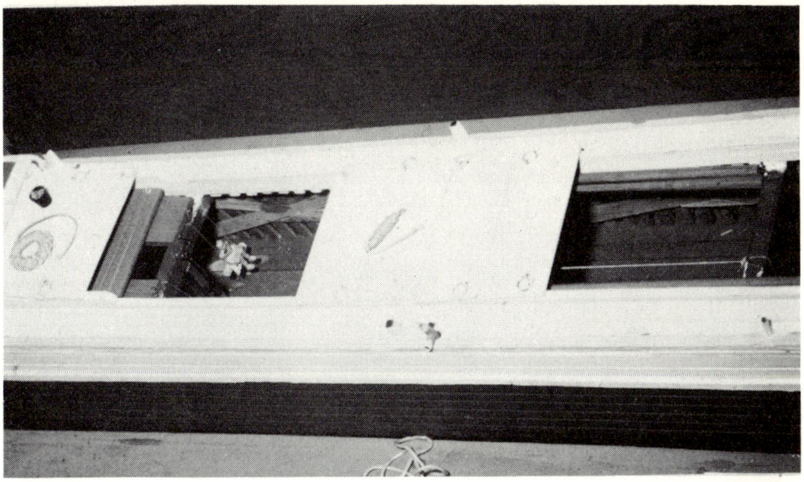

CLB 11 M. B. WAKEFIELD

During the run back to Honesdale Mr. Weber's father would hang a swing in the empty hold for the youngsters to play on. In the right hand open hatch area a clothesline may be seen tied to a support.

aboard a canalboat had two bunks along one side and one on the other. Each bunk was simply a straw bag with sheets and blankets over it. The closet for dishes was over the bunks. A drop-leaf table, that looked like part of the wall when not in use, and a small coal and wood stove (management frowned on the burning of coal on board the boats) completed the luxuries on board. CLB 10 and CLB 11

Frequently the cabin would be so crowded that the stove would be on deck. R 42. On many occasions necessity would find the boat operated by only two

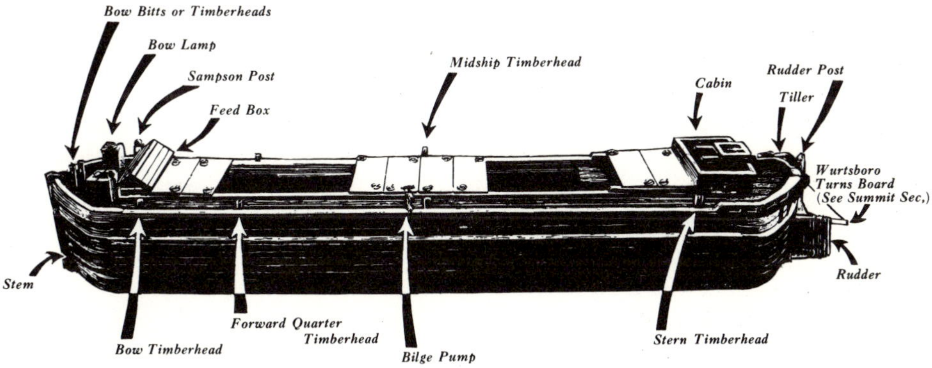

D & H CANAL BOAT CIRCA 1890

men, one who did the driving on the towpath while the other manned the tiller. The man who happened to be on the tiller at mealtime did the cooking. As Alvin F. Harlow tells it, "He would lift the potlid and take a look at the beans, run about forty feet to the stern and give the rudder a twist, trot forward to the stove, put on the coffee and break an egg into it, run back and give the tiller another swing, possibly take out the tiller handle until a bridge was passed, put it in place again and dash forward to snatch off the coffee pot, which was boiling over. When he had gulped down his hasty meal, he would leap to the towpath and drive while the other man ate. One old-timer tells us that 'if it rained, an umbrella must be held over the stove' but how the steersman could do this under the circumstances described, it is hard to guess." 4

CLB 10 Collection of M. B. WAKEFIELD

The David McClure family awaiting a permit to pass up the canal to Honesdale for a load of coal. Note the recently caulked seams.

The canal adjacent to the locks was always wider than the average width of the rest of the canal. This permitted boats to tie up when darkness would overtake the boaters. The locks originally closed at sunset but as business increased they were kept open till 10 p. m. Business gradually slackened off in the eighties until by 1890 the hours were changed to 6 a. m. to 6 p. m. The ruling prohibiting boating on the Sabbath was continued throughout the entire lifetime of the canal.

Canallers had to be up early in the morning to be ready for lock operations, which generally started at five o'clock. Once underway the horses or mules plodded along about a foot from the edge of the canal. When a narrow place on the towpath was encountered, one horse dropped behind the other. The harness for both animals, sharing a common pully, permitted pulling in-line or abreast. CLB 12. This device was located just above the shins so as not to hinder their movements. The canaller always had his best horse in front when *in-line* pulling was necessary.

Horses became the identifying symbol for some boat operators. For example, the father of DuBois Weber was known as the *Grey Liner,* because of the six grey-white horses he used to pull the three boats he had in operation.

As time went by more and more mules were used on the canal. They were tougher than horses, cheaper to maintain, and even had the virtue of dying rapidly if they took sick. CLB 13

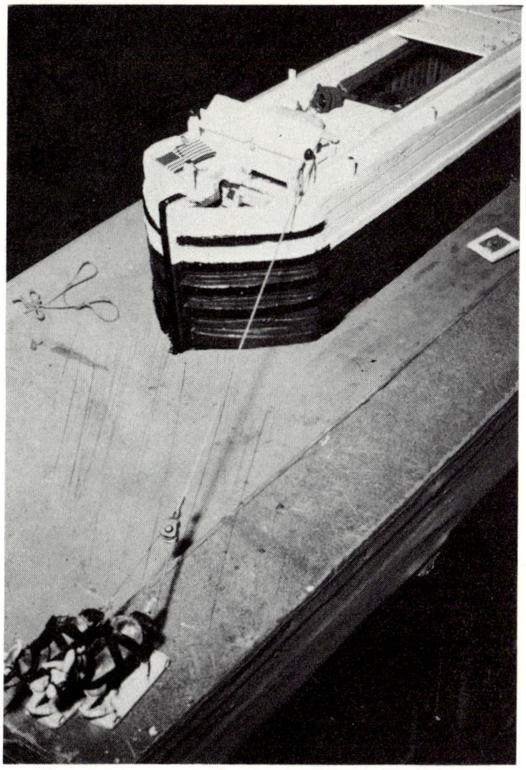

CLB 12 Collection of M. B. WAKEFIELD

The pully device, (out of scale) is shown in the Weber canal boat model display. The towrope is connected to the port Sampson Post in back of which is the feed box.

The stamina of the mules has been borne out by their ability to pull canalboats while their eyes were closed and they were virtually asleep on their feet.

John A. McGinnis of Honesdale related to the author on one occasion, "I've seen fellers walking between the mules carrying hay; the mules would be chewing on it and their eyes would be shut." This frequently happened when boatmen would be pushing hard night and day to make Eddyville before a sudden freeze at the end of the boating season would trap them out in the countryside. When a boat was caught under such circumstances the boatman simply wrote the load off as a total loss. Farmers from all points as well as villagers descended on the immobilized boat and made off with the coal.

The canal company had men with scow brigs to help boatmen who were caught by the ice. **5**

John A. McGinnis recalled being "froze-in" near Whittaker's Lock when the temperature dropped suddenly one night in December. The scow brig captain "had only one horse so he asked my father to hitch his team on to help. Well, we got through that time but I remember one year when they didn't get half the boats down because of the sudden freeze." If a threat of ice on the canal appeared, the locks would be kept open all night.

The Port Jervis Evening Gazette of January 6, 1870 reported, "About 300 canal boats are frozen in on the line of the Delaware and Hudson Canal, with cargoes aggregating 37,000 tons. There are 9,000 tons on the Rondout docks and frozen in on boats at that place are 4,500 tons more." R 41

There were other freight boats on the canal, known as *outside boats,* for which the D. and H. was not directly responsible. These boat companies transported a wide range of products and goods. The largest of the outside outfits was the Pennsylvania Coal Company (see Lackawaxen Section) which operated an extensive fleet of coal boats between Hawley, Pennsylvania and their company town, Port Ewen, on the Hudson River.

CLB 13 Courtesy of ELLENVILLE PUBLIC LIBRARY

Frequently canalboat operators would use one horse and one mule, keeping the horse on the canal side for leading purposes when in single file. This scene is about one quarter of a mile from Tom Quigley's lock below Ellenville.

The Cox Brothers at Ellenville operated six boats transporting cordwood to brickyards on the Hudson River. The Merchant Tanner Line, also at Ellenville, operated six boats in merchandise and local freight from canal points to New York City. CLB 14. James Lounsbury and Henry Green at Middleport operated four boats shipping cordwood to brickyards on the Hudson. Then there was A. J. Barnhart at High Falls who operated two boats shipping cement to New York City along with the A. J. Snyder Cement Company at Lawrenceville who operated two boats to New York and other points. CLB 15

The Lawrenceville and New York Cement Company at Lawrenceville operated two boats to New York along with the Rosendale Cement Company which operated three boats in bulk cement to Eddyville. The New York and Rock Lock Cement Company at Rock Lock operated ten boats to the City. The Maxwell Stone Company of Pond Eddy ran several specially built stone-carrying boats.

The powder boat of the Laffren Powder Company always gave the canallers some jittery moments until it was safely passed. This boat operated from canal ports to New York City.

Wilbur and Patmore regularly ran thirty-six deck canalboats for the transportation of merchandise from New York to Honesdale without trans-shipment, having a fleet of steamboats on the Hudson River solely employed in towing their freight boats from the foot of Jay Street, New York City, to Rondout.

Shortly after the opening of the canal, packets for passenger travel appeared. Although received with considerable enthusiasm, patronage fell off steadily over the years. Poor accommodations seem to have been the reason for the lack of business. In 1850 the *Fashion* came along to remedy the situation.

It was on June 12, 1850, that the beautifully modeled and handsomely equipped *Fashion* arrived in Honesdale for service to Lackawaxen to make connections with the recently completed New York, Lake Erie and Western Railway. The *Fashion,* during the season of 1850, left Honesdale every morning, Sunday excepted, arriving at Lackawaxen immediately on the arrival of the train going west, which left New York at half past six in the morning, and which was due at the mouth of the Lackawaxen at quarter to two in the afternoon.

The late T. J. Ham reported, "The graceful craft was built in Rochester, New York, and was a twin of the first class Erie Canal passenger packets. She was entirely new, as was the furniture, bedding, and all other equipments. New accommodations, beside ample dining and lunching facilities, permitted the lodging of sixty persons and, as a choice stock of refreshments suited to every palate was always politely served on board, it may be imagined that a trip on the boat could be made a very enjoyable one indeed.

"Before the *Fashion* commenced her regular scheduled trips between Honesdale and Lackawaxen, a large party of townfolk were treated to an excursion over the line.

"The general spirit of good nature was heightened from time to time when a bridge hove in sight. These structures were not built to accommodate the

CLB 14 Collection of M. B. WAKEFIELD

The "Ulster Queen", a freight boat owned by outside boat operators Hunt and Donaldson, is shown just below Ellenville in the last years of canal operation.

CLB 15 Courtesy of ELLENVILLE PUBLIC LIBRARY

Loaded cement boats moved directly to New York City in Cornell Steamboat Company tows. DuBois Weber recalled seeing the "Minnie S" named for Mrs. Snyder of the A. J. Snyder Cement Company, discharging a 1000 barrel cargo on the Harlem River in New York. Here the "Minnie S" is shown in the Rondout just below Eddyville.

passage under them of high decked boats with much room to spare, and when the *Fashion,* with most of her jolly excursionists on deck, came along, the warning cry 'low bridge' caused a general ducking of heads as the only alternative to a ducking in the canal. The wags took advantage of this fact and amused themselves every now and then by shouting 'low bridge' when no such obstruction was ever in sight. Down would go the heads of the crowds instinctively, a general mixture of hats and bonnets would ensue, and then, after the imaginary danger was over, a hearty good laugh over the joke put everybody in better humor than ever." **6**

The *Fashion's* career was doomed to a sudden and dramatic end. On the night of the 25th of April, 1851, a disastrous fire broke out which claimed as one of its victims the popular *Fashion.* The packet, along with other boats lying in the lateral slip then known as Wilbur and Patmore's Basin, was awaiting the delayed opening of navigation on the canal.

She was burned to the water's edge.

A boating delay and a fire banished forever such fancy trappings from the waters of the Delaware and Hudson Canal.

4

The Lackawaxen Section

The great conflagration that swept Honesdale on that memorable night in April of 1851 started at half past ten o'clock in the buildings of Capt. Edward Murray.

The burnt district extended from the initial starting point to Second Street and thence down to Sixth. Every building with but two or three exceptions between Second and the canal was gutted by the holocaust. Among the heavy losers in the fire, which was undoubtedly of incendiary origin, was Wilbur and Patmore, as noted earlier. Their warehouse was destroyed and the lumberyard reduced to ashes along with the costly freight boat *Daniel Webster.* L 1

As a result of the great fire, Honesdale regressed temporarily to the developmental period of September 1827 when Samuel Kimble sold all that portion of the present borough of Honesdale lying south of the middle of Central Park to Maurice Wurts of Philadelphia. For this, Kimble received a thousand dollars. One month later Wurts sold the property to the Delaware and Hudson Canal Company for four thousand dollars. This was the beginning of the little settlement at the Forks of the Dyberry that grew into the bustling community of Honesdale. [1]

This community was named for the man who did so much in making the Delaware and Hudson Canal Company become a reality and who by coincidence died at the age of seventy-one, just ten days after the '51 conflagration. Known unofficially as Honesdale, it wasn't until after the visitation of the noted author Washington Irving along with Philip Hone in 1841 that it was officially named Honesdale.

The first public eating and lodging house in Honesdale was the Wayne County House, built in 1827 by Charles Forbes. This hostelry was patronized by persons traveling by stage between Carbondale, Milford, and Narrowsburg,

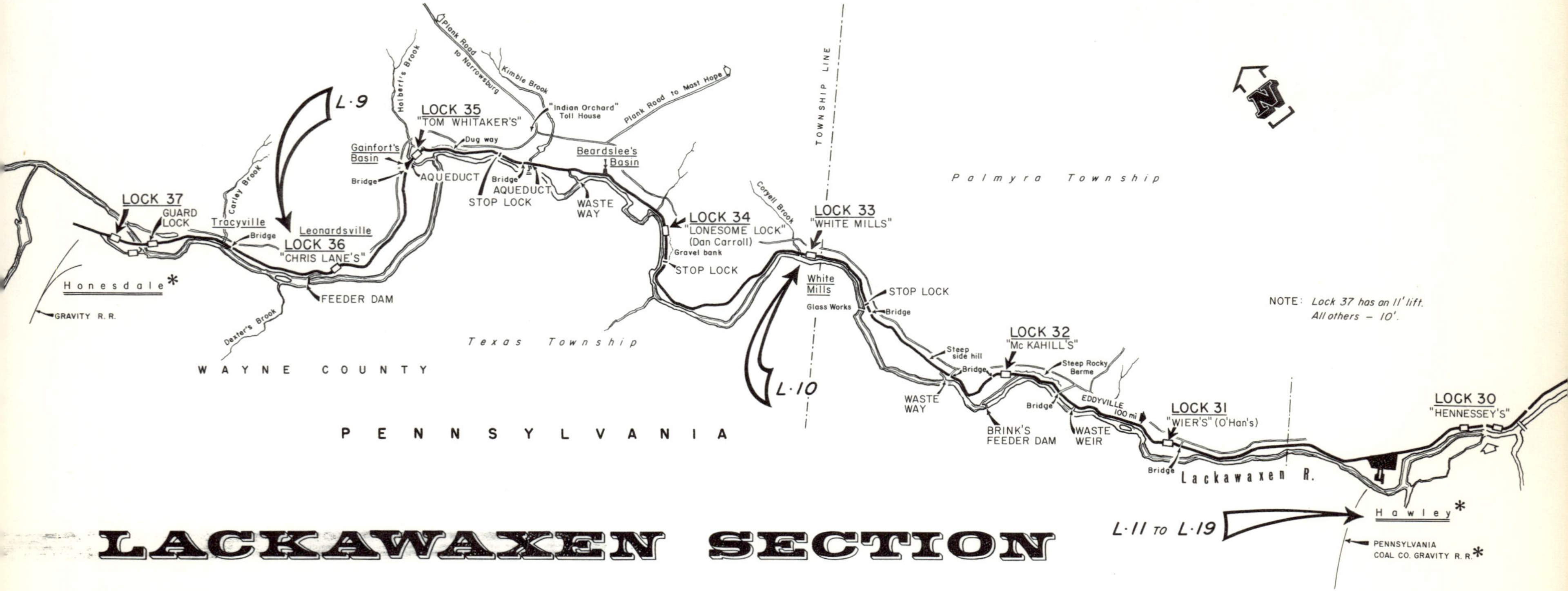

LACKAWAXEN SECTION

Honesdale to Hawley

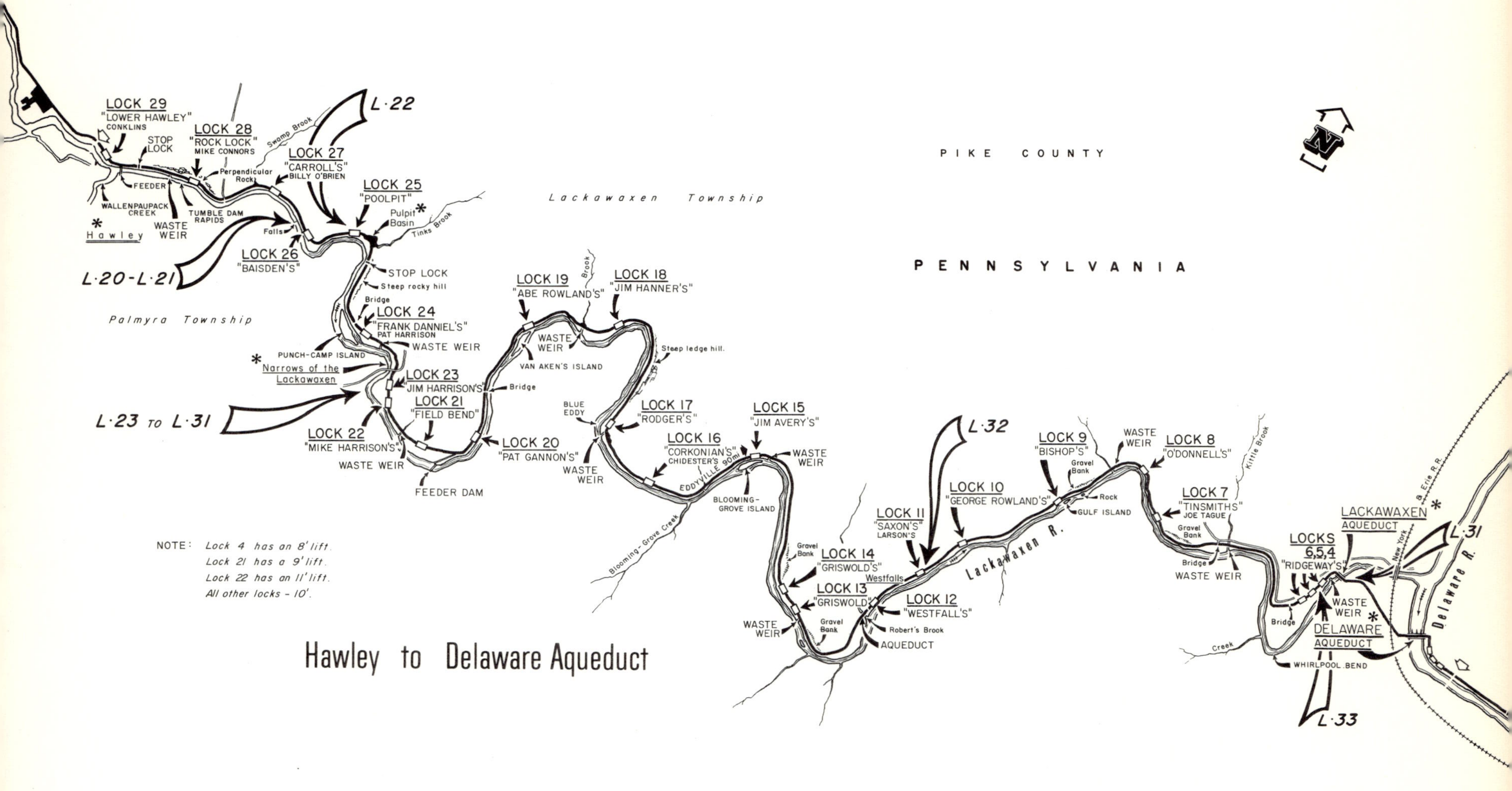

NOTE: *Lock 4 has an 8' lift.*
Lock 21 has a 9' lift.
Lock 22 has an 11' lift.
All other locks - 10'.

Hawley to Delaware Aqueduct

L 1 Collection of M. B. WAKEFIELD

The bridge crossing the slip leading to Wilbur and Patmore's Basin where in 1851 the great fire burned the packet boat "Fashion" to the water line.

N. Y. over the plank road. Here Philip Hone stayed when visiting the construction sites in his capacity as the first president of the company.

The Washington Irving trip up the canal was noted by Hone in his famous diary. "Their whole voyage was one of mirth and good cheer. They took inconveniences of the small canal boat, making their beds on the hard planks, eating in primitive fashion and traveling three miles an hour." To record forever the distinguished guest's visit, the rock mass brooding over Honesdale was affectionately named Irving Cliff.

Looming over Honesdale on the opposite side of town were the sprawling piles of coal. L 2. Bisecting the hill was Plane 13, the steepest on the line, which led to the depot and offices of the D. and H. Canal Company. The cars, classified as to type of coal, moved to the loading basins where the empty coal boats were moored. L 3. The storage piles were built up during the winter months when the canal was frozen over, to be drawn upon about April 1st when boating commenced. The cars were pushed by hand to a turntable near the engine house where the tracks radiated out like spokes to the extremities of the piles. L 4

The planes and spiderweb-like pattern of the Gravity tracks held endless fascination for young boys down through the years. A typical youngster was Edwin B. Callaway who relates, "One of the thrills of my early days was a ride to Gill's Latches from the foot of Number 13 Gravity Plane in an empty coal car. [2] The late Issac H. Ball, who was a dock boss, asked young Benjamin and me if we would like to take a ride to Gill's Latches? Continuing, he said we could return on a loaded coal train. Both of us were overjoyed at the invitation and climbed into the dirty, coal-black cars. Mr. Ball told us to get up front and 'hang on' he standing back of us. It was time for the footman at the bottom of the plane, after he had placed the 'sling' (a chain with hooks at either end) into links in the cable of the plane and the car, to pull the signal wire, which rang a bell in the stationary engine room at the 'head' of the plane to indicate to the engineer the type of cars to be drawn up the plane. One pull was for coal cars, two pulls freight and three pulls passenger cars. Coal cars were being drawn up before we climbed into a car, and other coal carrying cars would follow us. Were we to ascend the plane with only one pull of the wire? No, the footman pulled the wire three times and our trip of six cars started gently as a passenger car filled with people. The spinning cable wheels and the links of the second cable going in the opposite direction, returning to the foot of the plane for another trip, was interesting to watch. Best of all, when our trip of cars reached the top of the plane, after the car runner had alighted the car and thrown the 'sling' unhooking our car from the cable, we looked down in the boiler room and there stood the late John Rode of Willow Ave., engineer, and turning our way, the smile on his face as we appeared, will never be forgotten by the writer." **3** G 17

Some Honesdale families found security in all phases of the Delaware and Hudson Company operations. One such example was the Baird family, of whom Mr. Edward F. Baird saw service in three different modes of transportation

L 2 Courtesy of J. VANCE HUNT

This overview of Honesdale clearly shows the towering piles of coal on the hill overlooking the boat basin. Running directly up the hill on the right is light Plane 13 while cutting diagonally down the hill is the Union Plane leading to the Erie Pockets. Fifth Street runs up to the left-hand edge of the picture to what is known as the Terrace St. Bridge. The Lackawaxen River skirts the foreground edge.

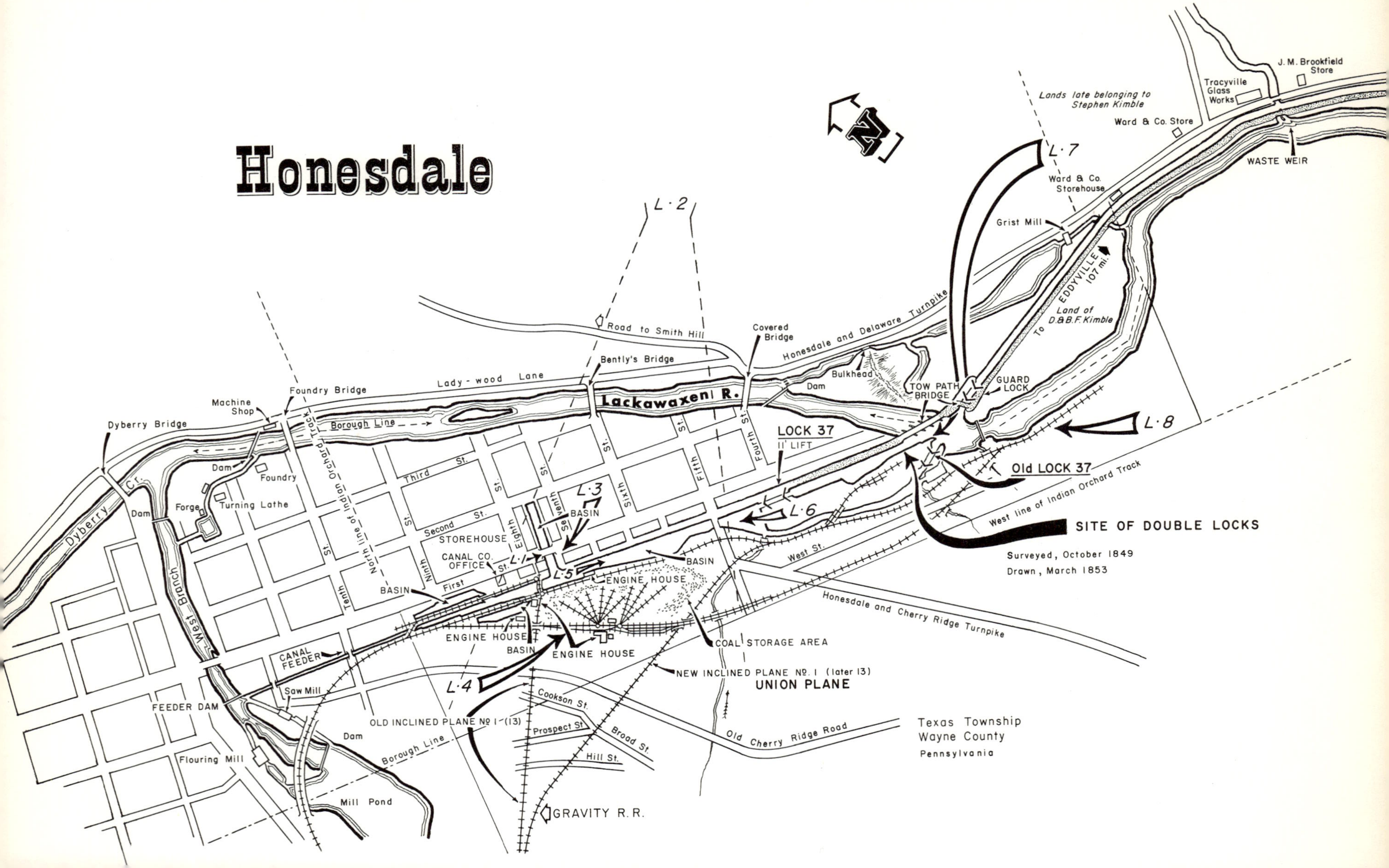

Honesdale
L·2
L·7
L·8
L·3
L·1
L·5
L·6
L·4
Lands late belonging to Stephen Kimble
J. M. Brookfield Store
Tracyville Glass Works
Ward & Co. Store
WASTE WEIR
Ward & Co. Storehouse
Grist Mill
To EDDYVILLE 107 mi.
Land of D.&B.F. Kimble
Honesdale and Delaware Turnpike
Road to Smith Hill
Covered Bridge
Bentlys Bridge
Lady-wood Lane
Foundry Bridge
Machine Shop
Dyberry Bridge
Lackawaxen R.
Borough Line
Bulkhead
Dam
TOW PATH BRIDGE
GUARD LOCK
LOCK 37
11' LIFT
Old LOCK 37
West line of Indian Orchard Track
SITE OF DOUBLE LOCKS
Surveyed, October 1849
Drawn, March 1853
Fourth St.
Fifth St.
Sixth St.
Seventh St.
Eighth St.
Ninth St.
Tenth St.
Third St.
Second St.
First St.
North line of Indian Orchard Track
Dam
Foundry
Forge
Turning Lathe
Dyberry Cr.
West Branch
STOREHOUSE
CANAL CO. OFFICE
BASIN
BASIN
BASIN
BASIN
ENGINE HOUSE
ENGINE HOUSE
ENGINE HOUSE
West St.
Honesdale and Cherry Ridge Turnpike
COAL STORAGE AREA
NEW INCLINED PLANE No. 1 (later 13)
UNION PLANE
CANAL FEEDER
Saw Mill
FEEDER DAM
Flouring Mill
Mill Pond
OLD INCLINED PLANE No 1 (13)
Cookson St.
Prospect St.
Broad St.
Hill St.
Old Cherry Ridge Road
GRAVITY R.R.
Texas Township
Wayne County
Pennsylvania

L 3 Collection of JIM SHAUGHNESSY

Boats being loaded in the upper basin. The farthest boat is riding light while the next boat is loaded. The boat in the foreground is being loaded from the stern forward at the same time that the boards are going up (hipping) to increase its capacity. The middle boat in this 1898 picture has already been hipped.

during his more than 50 years in Delaware and Hudson employ. "As a boy he drove horses and mules along the canal; later he held several jobs on the Gravity; and for 31 years he ran locomotives on the steam railroad . . .

"In 1880, at the age of 15, Mr. Baird was given a job as switchman on the Gravity at 40 cents a day, by Foreman Issac Ball. It was his duty to switch the coal cars coming from Carbondale into the docks where they were emptied into canalboats . . .

"Through the winter months Mr. Baird had two duties; he ran

L 4 Courtesy of ELLENVILLE PUBLIC LIBRARY

The loaded cars were rolled by hand out to the turntable where they were then shunted out to the ends of the coal piles for dumping.

L 5 Collection of M. B. WAKEFIELD

***Once the boats were loaded they were poled through the Honesdale Basin to the double locks #37 and 38. Two men, one on each side of the boat, would push the long pole into the bottom of the basin and 'walk the boat' down under the Terrace St. Bridge shown in background. Many of these buildings still stand today* (1965). (*See line drawing opposite Foreword*).**

errands for officials and took care of the torches used for illuminating the yard and storage piles when the men worked after dark. Frequently the entire force was kept busy until 11 p. m. disposing of the coal brought over the Gravity during the day. The torches used were ball-shaped, with three wicks projecting from the top.

"After two years as a switchman, Mr. Baird became headman at the top of Plane 13 . . . It was his duty to disengage the cable as the cars broke over the top of plane, then make up trains of from 48 to 72 cars for the run to Waymart . . .

"After ten years as headman, Mr. Baird was made stationary fireman at the head of Plane 13. Reporting for work at 4 a. m., he had to get up steam on the nine boilers, arranged in nests of three. During the course of a busy day he wheeled in from the storage pile and shoveled into the three hungry fire-boxes from 8 to 12 tons of buckwheat coal. At this engine house there were

L 6 Collection of M. B. WAKEFIELD

Looking from the double locks up the canal basin under the Terrace St. Bridge. Mountains of coal may be seen sloping down in the upper left corner.

two engines, one for Plane 13, the other for the Union Plane coming up from the Erie Railroad coal pockets." [4]

The routine of running the Gravity and loading the boats was shattered during the Civil War year of 1863 when the coal heavers went on strike. Just one year later, however, on June 6, 1864, the all-time daily record was set when 64 boats carrying 7,579.4 tons of coal arrived at Rondout. Six years later the miners went on strike causing hardships among the boatmen and an economic crisis in Honesdale. The strike dragged on through May, June, and into July when the Carbon *Democrat* reported that . . . "The D. and H. Canal Company contemplates introducing Chinese coolies to do the work of mining in their coal fields." [5]

By September over three hundred boats were tied up at Honesdale waiting for cargo, when the notice came through on the 4th that the miners had resumed work. The company made compensation to the boatmen for their good conduct during the strike by paying to the boatmen who operated to the close of the season "an advance of fifteen cents per ton to be made on all coal shipped at Honesdale on and after Monday, September 6th, making the rate of freight from that date, one dollar and fifteen cents per ton on coal carried from

Honesdale to Rondout, and in proportion for shorter distances." [6] For the balance of a severely shortened boating season the canallers did fairly well.

The boats were propelled down the lower basin to the double locks #37 and 38. L 5 and L 6. The water level in this basin and the upper part of the Lackawaxen section was maintained by the establishment of a series of feeder ponds and a major feeder from the West Branch of the Dyberry. [7]

Sometimes altercations developed if a boatman hired a Negro as a teamster or if one boatman thought he was getting a poorer grade of coal than the other fellow. Mr. F. J. Lewis recalled how a boatman, disgruntled over the quality of his coal, "tried to ram our boat in one of the double locks in Honesdale. My father was driving the team, and my mother was attending the tiller and steering the boat. The towline of this boatman hit the steering wheel and my mother at the same time. She saved the day by cutting the towline with a hatchet. This caused the boatman's team to go into a huddle and his boat ran aground. The D. and H. officials at Honesdale then took a hand in the matter. Just prior to this affair several boatmen, unknown, tried to waylay my father one night. He hid from them. They then tried to destroy the cabin. When they tried to open the cabin door, my mother shot at them. That ended the skirmish." [8]

Double locks #37 and 38 at the lower end of the Honesdale boat basin. The only double locks on the canal were located just above the still water crossing of the Lackawaxen River as shown in L 8.

L 7

L 8 Collection of M. B. WAKEFIELD

The towpath bridge across the Lackawaxen below the double locks shown at left. The boats returned to the canal in back of the barn at right. The dam in the foreground was maintained by the canal company for the purpose of creating a still-water pool for the safe crossing of boats. Note the covered bridge in the right background.

Double locks #37 and 38 were the only double locks on the canal, below which the towpath crossed the Lackawaxen to the guard lock. L 7 and L 8. This towpath bridge was for the convenience of the towing mules and at times of freshet required the special services of Larry Weidner. Mr. Weidner was hired by the company for just such occasions and if any damage resulted, the company paid the repairs. It seems the canalboat captains did a better job than old Larry. John A. McGinnis commented, "We used to put them across and wouldn't touch going in, but everytime he'd put them across, they'd hit."

At Leonardsville was located the boat works of Christopher C. Lane for which Lock #36 received its popular name. L 9. Downstream at White Mills was located the extensive glass works of Christian Dorflinger. It was established about 1870 and was by 1880 the largest glass manufacturing works in the county. L 10

One of the important communities on the route of the canal was Hawley, the terminal transfer point for the Pennsylvania Coal Company Gravity Rail-

L 9 M. B. WAKEFIELD

Tree-grown Lock 36 at Leonardsville was also known as 'Chris Lanes' adjacent to the famous boat works operated by Mr. Lane and apprenticed by Levi Barker. Later Barker went to Hawley where he amassed a fortune building boats.

road, and a vital boat building center. It was the hamlet of Paupack Eddy when the D. and H. Canal went through and it wasn't until 1837 that Joseph Atkinson erected and opened the first tavern which later became a hotel. Also in 1837 the first post office was established called Paupack Eddy. Later it was changed to Hawleysburgh, only to be changed again in 1851 to Hawley in honor of Irad Hawley, the first president of the Pennsylvania Coal Company.

Mr. Levi Barker, who was employed as a boat builder at the Christopher Lane Boat Works, saw the advantages of Hawley as a boat building location and in 1849 left the Lane Works and erected a boat yard at Hawley. He immediately accepted a contract from the Pennsylvania Coal Co. for twenty-five of the large deck-type boats of a capacity of 140 tons. For almost 40 years Mr. Barker continued in boat building and repair work and during that time it is estimated that at least six hundred boats were constructed. L 11

The Pennsylvania Coal Company's Gravity Railroad from Port Griffith to Hawley was completed in the spring of 1850, and the first coal sent forward in May of the same year. In the following season 316,017 tons were shipped forward which was about three-fourths the quantity shipped by the D. and H. Co. during the same period of time.

The Pennsylvania Coal Company Gravity had twelve ascending planes on the loaded track between Port Griffith and Hawley and ten ascending planes on the light or empty tracks on the return trip. The longest descending plane on the loaded track was a little over fourteen miles long and the longest on the light track over twenty miles long. There were twenty-two powerhouses in all, the first on the loaded track at Port

ELAWARE & HUDSON CANAL COMPANY.

CANAL DEPARTMENT.

Collector's Office, Hawley Pa July 29, 1889

No. 917

... to points named below, under and subject to the ... and By-Laws established for the government of the Ca...

DESTINATION	MILES	ARTICLES	WEIGHT	RATE	TOLL
White Mills	14	10 Cord wood			2.16

W. A. [illegible] Collector.

L 10 Collection of M. B. WAKEFIELD

Permit for boat 917 to move to White Mills with a load of cordwood, possibly consigned to the Dorflinger works. Since the boat passed through the collector's office at Hawley the load was picked up somewhere between the Narrows and Rowlands, 12 miles away.

Griffith and number twelve, the last, also on the loaded track.

The first plane for return of empty cars was number thirteen, and began on the right bank of the Lackawaxen. Most of the planes were operated by steam, but numbers thirteen, fourteen, fifteen and sixteen used water wheels.

L 11 Courtesy of J. VANCE HUNT

The 'high works' at the Hawley Boat Basin. The dry docks and boat yard of Levi Barker are to the right out of the picture.

The company had over nine hundred coal cars in operation besides cars for moving other commodities, and that number was being increased daily at the rate of six to eight cars which were turned out by the car shops at Hawley. The coal car was an oblong box, ten and one-half feet long by three and one-half feet wide. "The car contained between six and seven tons of coal. The car itself weighing not more than five thousand pounds. After weighing the cars, they were hauled to the road crossing by an old calico horse in charge of Daddy Shelp, a distance of probably three hundred feet on the main track and when the track was filled with loaded cars to rear end of coal pockets, an engine would be attached and the loaded cars drawn slowly over the scales, each car being weighed as it passed, the weight and size of coal announced by the weigh master and a book record made by the clerk, who would also get the car number. When such weighing was completed, the cars would be billed or manifested to whomsoever and wheresoever orders were to be filled." [10]

Grease or oil pans were placed along the track four miles apart, with a brush attached to either end of an iron rod, held in place by springs with an upright piece of sufficient height to strike the axle and thus dip the brush onto the brass or the part of the axle bearing on it. Less than a forth of the cars were equipped with brakes, and the men would carry six or eight puddle sticks, as they were called, to be used when handling cars without brakes.

These sticks were shaped like a potato masher with a hole in the small end through which a wire was run in order to carry them. The brakes were a series of levers on the outside of the car and one on the end which the brakeman pressed down with his weight and held there by a rachet.

The railroad from number six plane, Dunmore to Hawley and return, was operated by twenty train crews. Each crew consisted of three men, a car-runner and two assistants. The train crews lived along the line at various places and

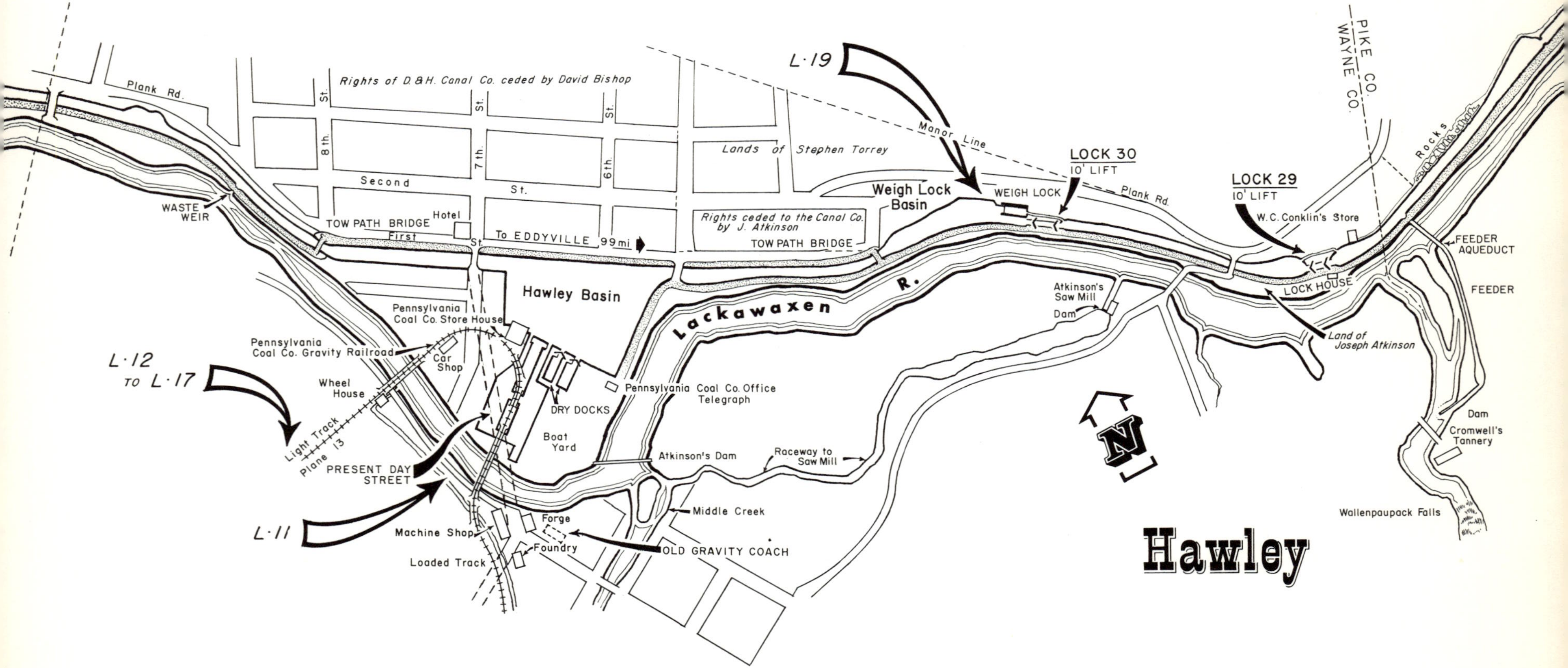
Hawley
L·19
Rights of D.&H. Canal Co. ceded by David Bishop
Plank Rd.
8th. St.
7th. St.
6th. St.
Second St.
Lands of Stephen Torrey
Manor Line
LOCK 30
10' LIFT
LOCK 29
10' LIFT
PIKE CO.
WAYNE CO.
Rocks
Weigh Lock Basin
WEIGH LOCK
W.C. Conklin's Store
WASTE WEIR
TOW PATH BRIDGE
Hotel
First St.
To EDDYVILLE 99 mi
Rights ceded to the Canal Co. by J. Atkinson
TOW PATH BRIDGE
FEEDER AQUEDUCT
FEEDER
LOCK HOUSE
Hawley Basin
Lackawaxen R.
Atkinson's Saw Mill
Dam
Land of Joseph Atkinson
Pennsylvania Coal Co. Store House
Pennsylvania Coal Co. Gravity Railroad
Car Shop
L·12 TO L·17
Wheel House
Light Track
Plane 13
PRESENT DAY STREET
DRY DOCKS
Boat Yard
Pennsylvania Coal Co. Office Telegraph
Atkinson's Dam
Raceway to Saw Mill
Dam
Cromwell's Tannery
L·11
Machine Shop
Forge
Foundry
Loaded Track
Middle Creek
OLD GRAVITY COACH
Wallenpaupack Falls
N

a day's work commenced by starting the train, either loaded or light, from their places of abode in the morning, and running to the end of the line and back to their homes before dark. The railroad was operated six days a week from six o'clock in the morning until about sunset. L 12, L 13, L 14, L 15, L 16 and L 17

L 12 Courtesy of J. VANCE HUNT

An exceedingly long cut of light cars skirting the base of the Greenville Cliffs.

On August 31, 1847, the Delaware and Hudson Canal Company entered into a contract with the Wyoming Coal Association out of which later grew the Pennsylvania Coal Company. An agreement was concluded between the two companies by which the coal forwarded by the Pennsylvania Coal Company was sold to the D. and H. Company at its terminal at Hawley. Later, on July 29, 1851, a formal statement was signed directing that a fair price was to be determined for the use of the canal after completion of the enlargement projects. Implicit in this was the assumption that the enlarged canal could move coal far more cheaply. One-half of the difference was to be charged to the Pennsylvania Coal Co. in addition to its regular tolls. This first agreement was the only one ever completed between the two companies and it was never enforced. [9]

L 13 Courtesy of J. VANCE HUNT

The rare instance at Plane #5 where the light track crosses over the loaded. The loaded track, shown here, is climbing out of Stafford Meadow.

On the first of May, 1852, the agreement was terminated on the grounds that the said agreement bore upon the Pennsylvania Coal Company's interests unequally.

Immediately differences of opinion arose as to the rightful interpretation of the original contract, and the question was referred

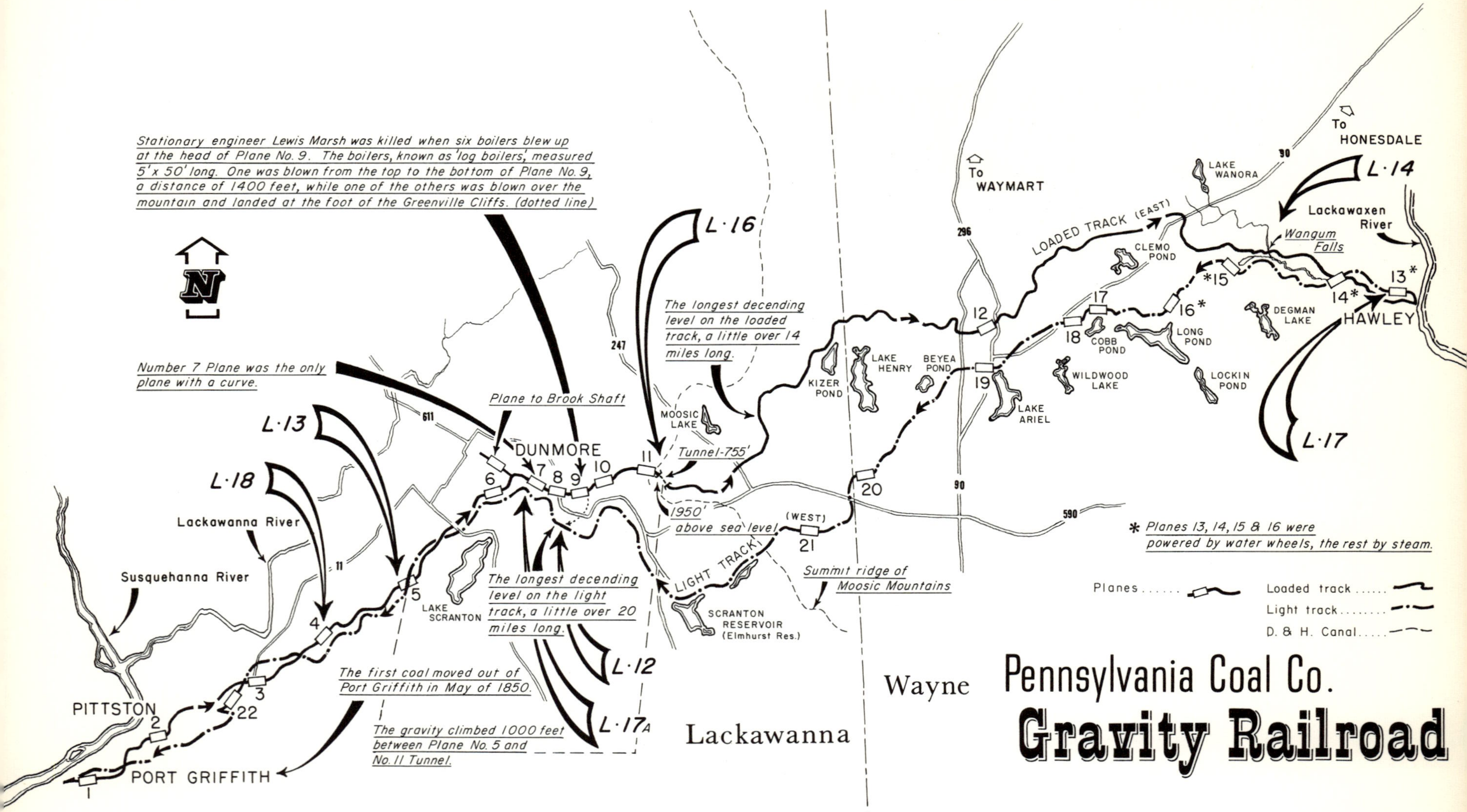

Pennsylvania Coal Co.
Gravity Railroad
Stationary engineer Lewis Marsh was killed when six boilers blew up at the head of Plane No. 9. The boilers, known as 'log boilers,' measured 5' x 50' long. One was blown from the top to the bottom of Plane No. 9, a distance of 1400 feet, while one of the others was blown over the mountain and landed at the foot of the Greenville Cliffs. (dotted line)
Number 7 Plane was the only plane with a curve.
Plane to Brook Shaft
The longest decending level on the loaded track, a little over 14 miles long.
The longest decending level on the light track, a little over 20 miles long.
The first coal moved out of Port Griffith in May of 1850.
The gravity climbed 1000 feet between Plane No. 5 and No. 11 Tunnel.
Tunnel-755'
1950' above sea level
Summit ridge of Moosic Mountains
* Planes 13, 14, 15 & 16 were powered by water wheels, the rest by steam.
Planes
Loaded track
Light track
D. & H. Canal
N
L·13
L·18
L·16
L·12
L·17A
L·14
L·17
DUNMORE
PITTSTON
PORT GRIFFITH
HAWLEY
To WAYMART
To HONESDALE
LOADED TRACK (EAST)
LIGHT TRACK (WEST)
Lackawanna River
Susquehanna River
Lackawaxen River
Wangum Falls
LAKE SCRANTON
SCRANTON RESERVOIR (Elmhurst Res.)
MOOSIC LAKE
KIZER POND
LAKE HENRY
BEYEA POND
LAKE ARIEL
WILDWOOD LAKE
COBB POND
LONG POND
LOCKIN POND
CLEMO POND
LAKE WANORA
DEGMAN LAKE
Wayne
Lackawanna
1
2
3
4
5
6
7
8
9
10
11
12
13*
14*
*15
16*
17
18
19
20
21
22
611
247
11
296
90
590

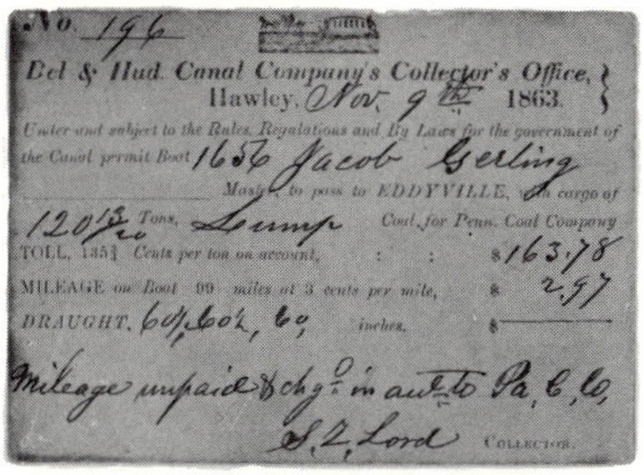
No. 196

Del & Hud. Canal Company's Collector's Office,
Hawley, Nov. 9th 1863.

Under and subject to the Rules, Regulations and By Laws for the government of the Canal permit Boat 1656 Jacob Gerling Master, to pass to EDDYVILLE, with cargo of 120 Tons, Lump Coal, for Penn. Coal Company

TOLL, 135¼ Cents per ton on account, $163.78

MILEAGE on Boat 99 miles at 3 cents per mile, $ 2.97

DRAUGHT, 60, 60½, 60 inches. $

Mileage unpaid & chgd in audt to Pa. C. Co.

S. L. Lord Collector.

L 15 Collection of M. B. WAKEFIELD

Pass issued to a loaded boat at Hawley for the run to Eddyville. Note memo on lower portion of card recording "Mileage unpaid and charged in audit to Pa. C, Co." This amount was never paid.

L 16 Courtesy of J. VANCE HUNT

The 755 foot tunnel under the summit ridge of the Moosic Mountains. This is on the loaded track just beyond Plane 11 at the start of a 14 mile descending level.

L 14 Courtesy of J. VANCE HUNT

The scenic Wangum Falls with a loaded train coasting down into Hawley. This section of right-of-way is now a dirt road.

by the D. and H. Canal Company to the courts of Pennsylvania and New York for decision.

During the years 1855 and 1856 the quantity of coal transported for the Pennsylvania Coal Company averaged 550,000 tons. In 1855 the tolls paid on the canal and railroad were just over $650,000 and in 1856 just under $585,000, a large proportion being received from the Pennsylvania Coal Co. Large as the tolls from that company were, there remained uncollected considerable additional sums. L 15

Meanwhile legal proceedings against the Pennsylvania Coal Co. were continuing over the unpaid balance of toll payments, creating an atmosphere of unpleasant relations and a determination to get another outlet to market their coal. Accordingly they concluded a contract with the New York, Lake Erie and Western Ry. to build a branch railroad from their main line at Lackawaxen to Hawley, only to have it thwarted by D. and H. Company pressure groups at the public hearings.

By 1860 the Pennsylvania Coal Company was successful in having a franchise granted to that company allowing it to construct and maintain a steam railroad to run the 16 miles to Lackawaxen.

L 17 Courtesy of J. VANCE HUNT

Passenger train about to leave Hawley on Plane #13. Pennsylvania Coal Co. high works show in the distance.

During 1862 and '63 the construction of the branch line was doggedly pushed ahead through rugged valley terrain to its completion in December 1863. On December 23, 1863, the first car of coal was loaded for movement and weighed by Frank Hardenberg. Passenger cars were attached to the first two coal trains that were run from Hawley to Lackawaxen with James Frantz at the throttle and Charlie Gorham covering as conductor.

L 17 A Courtesy of J. VANCE HUNT

The famous hanging rock that was one of the unusual geographical features along the right-of-way. It was blown off in 1884.

L 18 Courtesy of J. VANCE HUNT

Plane #4 on the loaded track at Rocky Glen. It has since been filled in.

James O'Connor of Scranton wrote of the day he " . . . saw the first locomotive that entered Hawley. It came up over the Erie's new branch from Lackawaxen in the summer of 1863. I was among the children who had just been let out of school. We assembled on the creek bank along River Street and looking across the stream feasted our eyes on the engine, which stood near the Pennsylvania Coal

Company's pockets. The machine was a wood burner with an expansive bell stack and was surrounded by a crowd of villagers." [11]

The gravity tracks were rearranged to accommodate new pockets which were built just east of the old Marble Hill Bridge. From that time on Hawley began to wane as a boat center. Only three boats continued to be loaded with the Pennsylvania Coal Company's product and those expressly for destination points between Hawley and Rosendale. Shipments for such points were weighed at the weigh lock adjacent to Lock #30 downstream from the Hawley Basin. L 19

This system of handling the coal at Hawley continued for about twenty years until the Erie and Wyoming Valley Railroad superceded the Pennsylvania Coal Company Gravity in 1885, after which the pockets at Hawley were abandoned as the coal cars were moved directly to the mines.

L 19
PENNSYLVANIA HISTORICAL & MUSEUM COMM.

The weigh lock at Hawley was used by boats consigned to points between Hawley and Eddyville. All other loads for external points were weighed at Eddyville.

Between Hawley and the Delaware Aqueduct the canal boatmen passed through twenty-six locks and wound endlessly in tandem with the tortured convolutions of the plunging Lackawaxen River.

Lock #26, just before the Pulpit Basin, was also known as Baisden's Basin or Baisdenville. "One day," recalls Miss Hattie Tuthill, "Grandfather and Grandmother Baisden were driving from Hawley to Baisdenville, about two miles south of Hawley, when somehow a towline, carelessly thrown, caught their horse and pulled horse, wagon and couple into the canal. In its frantic struggle the horse tangled in the fringe of Grandma Baisden's black silk shawl which I still have. Grandfather and Grandmother were saved by means of a pike pole thrust out to them but the horse drowned." [12]

Today (1965) the old blacksmith shop still stands, as well as the ruins of Lock #26. L 20 and L 21

Around the bend and another quarter of a mile downstream is Poolpit Basin, officially listed as Pulpit which DuBois Weber remembered, "They tried

L 20

The old blacksmith's shop at Baisden's, Lock #26. A portion of the lock wall shows in the immediate foreground.

L 21

The eminent historian, Henry G. Soete, at Baisden's Lock pointing out the depression in the ground which yielded one of the two snubbing posts now on exhibit at the Wayne County Historical Society Museum.

The 'bottomless' Pulpit Basin today shows the Lackawaxen back in its original bed. The towpath passed down on the right-hand bank which was part of the earthworks thrown up in 1828 to deflect the Lackawaxen River from its original bed.

L 22

M. B. WAKEFIELD

to tell me it had no bottom, but I knew better." L 22. The course of the river today is the canal right-of-way of yesteryear. It is today the way the surveyors and engineers found it, a wide deep pool fed by Tink's Brook. The river, making an abrupt right turn, was found to be too sharp for safe canal navigation with the result that the river was diverted into a new channel making the Pulpit a physical part of the canal proper. The depth of the basin was 12 to 16 feet deeper than the new river channel, a fact which contributed greatly to a serious break shortly before the completion of the canal in 1828.

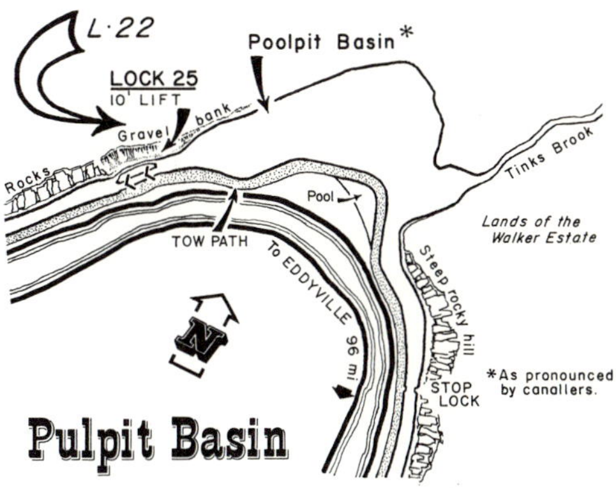

Diverting the waters of the Lackawaxen into a new channel caused great concern among the raftsmen of the upper reaches of that river.

Following is a partial loss accounting as set forth by Enos Woodward in 1828 and '29 to William Wurts, agent for the canal company.

1828 — Lost in the new channel near Pulpit
10,000 feet white pine valued at $10 per M $100.00

Taking out 5 rafts at $4.00 per raft 20.00

Rerafting same 5 rafts at $4.00 per raft 20.00

1829 — Lost from a raft at Pulpit 1000 feet white pine at
$9.00 per M 9.00

Lost from another raft 2000 feet white pine at
$9.00 per M 18.00

One raft white pine 20,000 feet at $9.00 per M
entirely lost 180.00

20,000 feet at $9.00 per M stuck at Pulpit and left in your charge. If the raft remains where it was stuck, I will take it back or not at the option of the company. They paying me in case I take it back $20.00 for the expense of taking and rerafting.

Extra hands and extra prices required in consequence
of the difficulty of running Pulpit - 14 rafts - 28.00

L 23 Courtesy of ELLENVILLE PUBLIC LIBRARY

The Narrows of the Lackawaxen measured 40 feet in width forcing the canal engineers to build a towering 30 foot dry retaining wall to contain the canal and towpath. The building in the background housed the W. W. Walker store.

Today the river has returned to its original channel and in so doing has completely erased the towpath and Lock #25.

Philip Hone, upon his return to New York City, recorded his impressions on a trip on the Delaware and Hudson Canal with the remarkable stonework at the *Narrows of the Lackawaxen* receiving particular attention.

The falls that once existed at this point were blasted out by the raftsmen in order to facilitate the safe passage of their rafts through the narrow defile. To permit the passage of the canal a shelf was blasted from the cliff and a towering dry stone wall erected to carry the towpath and canal channel. L 23, L 24 and L 25. From the opposite side of the river a large segment of this

L 24 Collection of M. B. WAKEFIELD

The massive dry wall at the Narrows looking downstream showing the canal side structures for Lock #23 in the distance. Pictures L 23 and L 24 were taken from approximately the same area on the opposite bank.

Today the dry wall shown in L 24 has vanished and in its stead towering pine trees have grown. By careful examination of the rock strata of the two pictures it is possible to determine where the towpath and wall once stood and to realize the engineering challenge it posed in 1827.

L 25 Collection of M. B. WAKEFIELD

L 26

A large section of the dry wall still stands supporting the towpath of Lock #23. The Narrows are at the left-hand edge of the picture.

L 28

The vertical stud remains of the wood lock lining at Lock #22.

wall may still be seen downstream just before Lock #23. L 26. Lock #23 is extremely well preserved with a rusting Model A where the lower gates would normally be. L 30. One of the unique features of this lock in the year 1965 is the overflow weir, still in remarkably good condition. L 27

Just a short walk reveals the remains of Lock #22 still containing most of the vertical studding to which the wood lining of the lock was once secured. L 28. Adjacent to Lock #22, lying across the towpath, is one of the original telegraph poles erected along the canal right-of-way in 1848. L 29 and L 29A. This seemingly insignificant pole is part of the first telegraph

The run-around overflow weir at Lock #23 just below the Narrows of the Lackawaxen.

L 27

M. B. WAKEFIELD

L 29

As late as 1954 some of the original telegraph line to be run still stood on the towpath west of the Narrows of the Lackawaxen. They have since vanished. One may still be seen (below) between Locks 22 and 23.

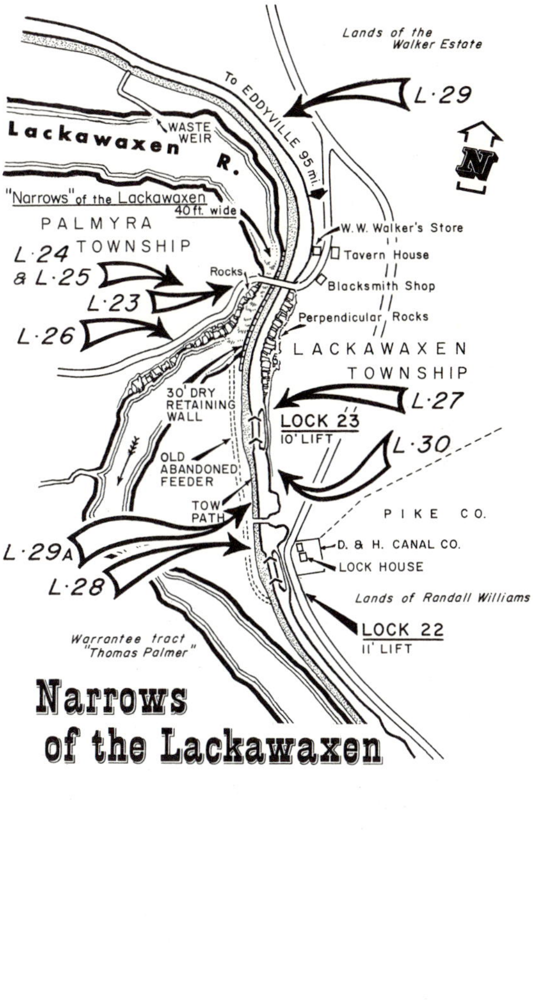

L 29 A M. B. WAKEFIELD

line commenced in August of 1847 by Ezra Cornell. The original line was along the public roads from New York through Harlem, White Plains, Peekskill, Newburgh, Goshen, [13] thence along the newly completed New York, Lake Erie & Western Railway tracks to the Lackawaxen, thence up the Delaware and Hudson Canal towpath to Honesdale, to proceed northwestward to Binghamton and Fredonia.

The Erie management, watching the experiment very closely, soon adapted the device to the operation of their trains, thus becoming the first railroad

L 30 **M. B. WAKEFIELD**

The passing basin below Lock #23 just beyond the Narrows of the Lackawaxen. DuBois Weber recalled that it was tended by a man named Jim Harrison, and the following lock by his brother Mike.

L 31 Collection of M. B. WAKEFIELD

This rare over-view of the Lackawaxen Aqueduct also shows in the background the three seldom photographed Ridgeway Locks. The apparent lack of water and the missing lower gate sheds and bridges would indicate the canal has been abandoned.

in the nation to do so. The Delaware and Hudson Company also quickly adapted the invention to their operation of the canal with offices at Eddyville, High Falls, and many other points.

This pioneer organization that the D. and H. Canal Company obliged with their right-of-way grew into the Western Union Telegraph Company.

For the remainder of the distance to the Delaware the highway is built on the canal route with but few exceptions. An elert eye will pick out waste weirs here and there or one side of a lock through which the road passes. L 32. It is a lonesome stretch even today so it can

L 32 M. B. WAKEFIELD

The highway up the Lackawaxen Valley follows directly in the canal right-of-way, passing through many of the locks, such as Lock #11, shown here. It was also known as Saxon's. DuBois Weber knew it as Larson's during the last years of boating.

L 33 Courtesy of PENNSYLVANIA HISTORICAL AND MUSEUM COMM.

A loaded coal boat passing out of Ridgeway Lock #5 into the basin before Lock #4 and the crossing of the Lackawaxen River Aqueduct.

well be imagined that the canaller felt some apprehension when darkness overtook him along this part of the canal.

He had cause for concern when we consider the release that appeared in the Port Jervis *Tri States Union* on May 3, 1878. "On Saturday night a number of poor boatmen were robbed by a company of seven tramps, operating between Hawley and Lackawaxen. The amount of their plunder is not known, but it cannot be very great, as this was the first trip of the season for most of the boatmen, who seldom have any work during the winter, and, as a class, start out poor in the spring. But the several small amounts that were taken, ranging from two to five dollars, are serious losses to men who had not any more to lose. One poor captain, who had just one dollar left to finish the trip with begged so piteously that the robbers returned it to him."

Following times like this the boatmen no doubt breathed a sigh of relief while locking through the twenty-eight foot drop of the three Ridgeway Locks just before the Lackawaxen Aqueduct. L 31 and L 33

Delaware & Lackawaxen Aqueducts

"Build the canal above the water."

So spoke John August Roebling while discussing the boating problems at the Delaware crossing in 1846 with the managers of the Delaware and Hudson Canal Company. The solution was as obvious and simple as that, at least to Roebling, who was then in the final phases of perfecting the principle of wire rope hangers for bridge suspension.

The problem at the Delaware River crossing was brought about back in 1764 by a man named Daniel Skinner. During a hitch as a sailor in the West Indies he had learned the value of pine lumber for masts and spars for ships.

Having a quantity of good pine on his land on the upper Delaware he put several sticks into the Delaware River to investigate the feasibility of floating them to Philadelphia. By following them in a canoe he quickly saw that before getting far along toward their destination the logs were aground on islands or stuck on rocks. He next put into the river six large shipmasts of equal length, through each end of which he cut a mortise of about four inches square, and into this he put what he called a spindle of white oak, to fit the mortise. In the end of this he inserted a pin to keep it from slipping.

"The lumber thus put together he called a raft, and to each end of it he pinned a small log cross-wise, and in the middle of this he fastened a pin, standing perpendicular about ten inches above the cross-log, on which he hung an oar fore and aft. It being thus rigged he hired a very tall Dutchman to go on the fore end, and with this raft arrived safely in Philadelphia, where he sold it at a good price". [1]

A classic **1890** *overview of the Delaware Aqueduct crossing showing a fast-moving coal train on the New York, Lake Erie and Western Railway. To the left of the dam is the old guard gate feeder. Note the wooden floating bridge in the lower left hand corner at the mouth of the Lackawaxen.*

DL 2 Collection of JAMES A. O'BOYLE

This was the first raft so constructed and run down river. Daniel Skinner was known forever after as "Lord High Admiral" of all the raftsmen on the Delaware. Before a new man could launch himself into a career as a raftsman he had to first present himself "at court" to Skinner with a bottle of spirits.

The raftsmen came to be the rulers of the Delaware, rulers over every rift and eddy with virtually nothing to stand in their way. That is, until 1827 when the Delaware and Hudson Canal Company built a small dam across the Delaware just below the confluence of the Lackawaxen River, to create an area of still water for the floating across of canal boats. DL 1 & 2

The altercations that developed between the boat operators and the raftsmen can well be imagined. The result was endless turmoil, loss, and litigation because of the recurring raft and canalboat mishaps.

The claims of the raftsmen, of both the Delaware and Lackawaxen Rivers, against the canal company over a typical period of three months totaled $8,792.27. Out of a total of 28 claims only 14 were paid. Some as high as $2,513.13 were ultimately paid. The Middle Creek Company claimed a loss of $1,917.50 and received payment. John R. Compton claimed $44.00 and settled for $20.00 while Zeke Reynolds really had a change of heart when he claimed $220.00 and settled for $5.00.[2]

DL 1 M. B. WAKEFIELD

Looking upstream from the Roebling Delaware Aqueduct showing the swell in the water caused by the remains of the company dam erected in 1827. The ice floe covered point of land in the upper left hand corner marks the Pennsylvania anchorage of the old rope ferry in use until the completion of the Roebling spans.

As the canal was originally built, the loaded boats dropped down through three locks, #3, 2 and 1 respectively, to the rope ferry crossing. On the New York side light boats moved out through a guard lock to the stilled pool of water above the dam. "A towpath was formed along the river edge a distance of ½ mile . . . to a point where a ferry was erected; by means of a pier stationed at the opposite side of the river composed of four foot square pine timbers locked together at the corners and the interior thoroughly filled with stones. The piers were twelve foot square at the base, about fifteen feet high and contracted to about seven foot square at the top. These piers supported the ends of a ferry rope two inches in diameter stretched across the river from pier to pier. By means of this rope a ferry scow was guided across the river as occasion demanded." [3]

The New York anchor of the dam abutted the towpath side of the guard lock masonry which, "stilled the water of the river back to a distance of about 1¼ miles forming a beautiful sheet of calm water during the dry season. A channel was provided for the passage of rafts and the height of water for the requirements of the canal regulated by means of brackets across such channel and removed when the volume of water in the river became excessive." [4] During times of severe drought, graveling the dam [5] was resorted to and when the water was excessively high the guard lock was closed and the canal fed through its paddle gates. This gate was the primary feeder for the Delaware section with only the Mongaup feeder, the Shinglekill and numerous small streams entering at different points acting as supplements.

When the water was at low mark the boatman would urge his horses to an extra burst of speed so as to establish sufficient headway to cause the boat to shoot across the river. This avoided the tedious process of being pulled across by rope.

"Many times the loaded boat crossing from the Pennsylvania side would pass over the river and enter the canal in New York before the horse and driver

DL 3 Courtesy of JAMES A. O'BOYLE

The Delaware Aqueduct from the downriver New York side. The old 1827 - 1848 canal is located in the foreground acting as a feeder with the guard lock out of sight behind the right hand abutment. The old dam shows under the aqueduct with the wooden bulkheads built to guide raftsmen between the abutments. Note water leaking out of the aqueduct on the left.

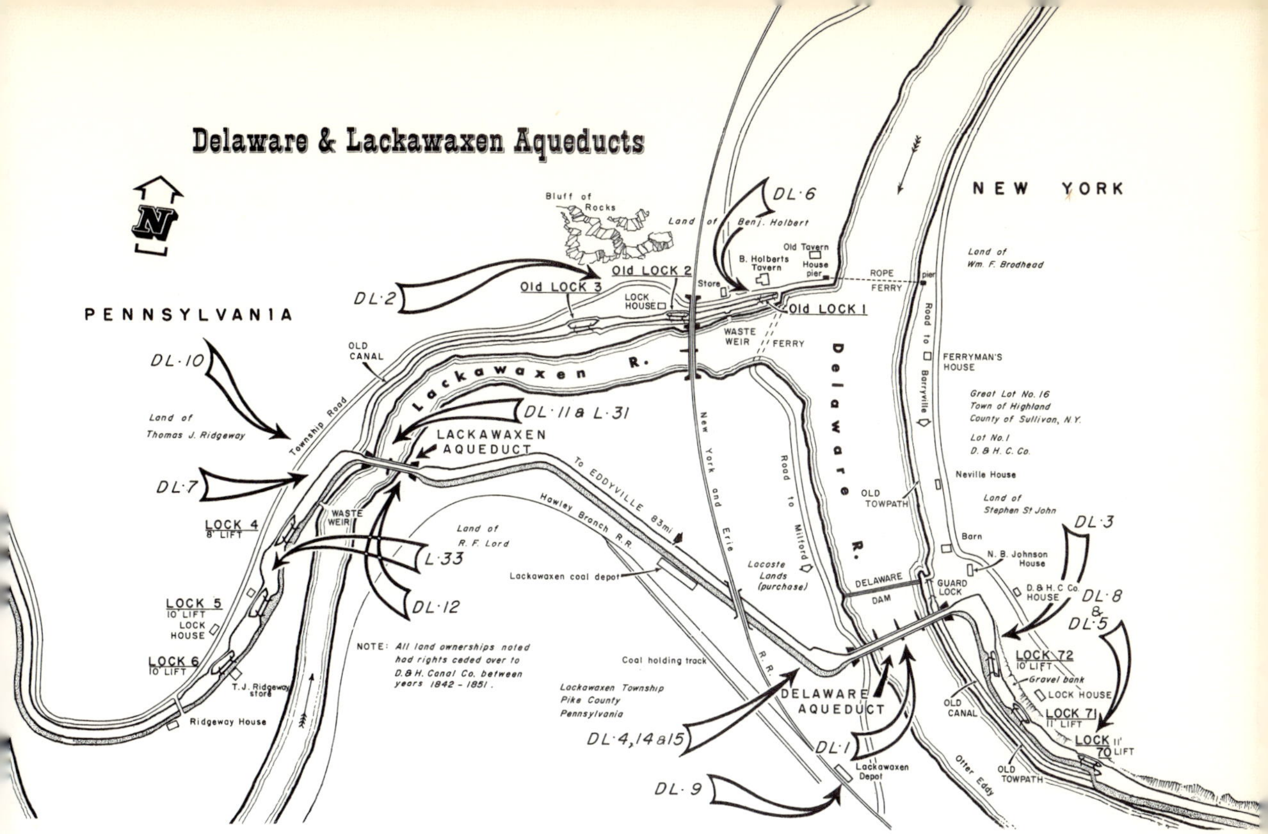

crossing by ferry would overtake it. When, however, the river was swollen by rains, the boats, horses and all must be crossed by the ferry, a slow, tedious . . . process requiring the strength of four and five men . . . on the rope. Even this was possible only at certain levels of water above which boats could not cross at all and the business of the canal suspended for the time." [6]

The action of the Delaware River, on the New York side, for the first mile or so of canal was a source of great expense to the canal company. High water and ice jams destroyed much of the towpath leaving vast deposits of mud and debris in the canal. DL 3

John Johnston has known such deposits to extend one full mile from the guard lock, in places nine to eleven feet deep with at least 100 laborers employed from the early part of January until late April for its removal.

As the tonnage on the canal increased so did the aggravation to the raftsmen. Considerable delays were encountered at the crossing even under good operating conditions, resulting in a loss of revenue to the company. The managers were becoming increasingly aware of the need for improvement at the Delaware crossing. On February 13, 1846, they authorized the construction of two aqueducts, one across the Delaware and one across the Lackawaxen.

On December 28, 1846, they received two plans for each of the aqueducts, one plan for a wooden structure and the other for a wire suspension. The two plans were referred to committee and on January 6, 1847, reported in favor of the wire suspension aqueduct. The main advantage of the wire structure, in their opinion, was that two fewer piers (three instead of five) would be necessary, thereby allowing more room for ice floes in time of flood. [7] Engineer Lord inspected Roebling's Pittsburgh suspension bridge in December of 1847 and returned with a favorable report.

Roebling was directed to start the project and to expedite it as rapidly as possible. The line of the canal was altered to accomodate the new crossing of the river which entailed the building of an additional aqueduct over the Lackawaxen one mile above its confluence with the Delaware.

It is interesting to speculate as to why two aqueducts were built at this point at a combined cost of $60,400. To have crossed at the site of the rope ferry would have involved the building of only one aqueduct. Also an aqueduct at this location would have been subjected to the ice floes and flood waters of only one river instead of two. Politics or problems in land acquistion may have been the reason.

"In the fall and winter of 1848 Roebling was building the Delaware aqueduct, stringing two 8½ inch cables over the saddles on top of the masonry piers. From these cables - using *wire rope hangers* for the first time - the canal trunk was suspended with a planked towpath on each side". [8] The aqueduct contained nearly eight thousand cubic yards of hydraulic cement masonry in its abutments and piers and with the extensions covered a distance of six hundred feet. Consisting of four spans, the width of its trunk at the water line was nineteen feet with a water depth of six feet. The diameter of its wire cable measure out to eight and one-half inches. [9] The maximum strength of each cable was nineteen hundred tons with the total weight of cables and anchor chains placed at four hundred and ninety thousand pounds.

By the end of the year 1848 the Delaware Aqueduct was completed and ready for navigation the following spring. DL 4. There was some apprehension in the crowds on hand when the first boat went across as to whether the bridge would stay up or not. After all, reasoned the expectant spectators, the water by itself weighed a great deal, let alone a loaded boat too. [10] Great cheers went up when the boat arrived safely on the New York side to pass on down through the newly built Locks #72, 71 and 70 respectively. On the Pennsylvania side Lackawaxen Locks #1, 2 and 3 were eliminated. DL 5 & DL 6

Even after the opening of the aqueduct the dam was still maintained. It was in fact heightened to 16 feet to increase the effectiveness of the stilled water area for canal feeder purposes. When rafts crossed over the dam the front end would, along with its bow oarsmen, disappear from the view of the stern oarsmen. The entire bow section would vanish under a swirl of white water with the result that the craft could not be controlled for the shot between the abutments. Accuracy on the approach to the dam was the deciding factor between success or disaster.

The canal passed under the New York, Lake Erie and Western Railway [11] tracks in the section between the Lackawaxen Aqueduct and the Delaware Aqueduct. Occasionally the passing of a snorting locomotive coincided with the passage of a canalboat with consequences of alarming proportions as noted in the May 10, 1887 issue of the Port Jervis *Evening Gazette.*

"Monday morning the captain and owner of a canal boat on the D. and H. Canal was proceeding on his slow but certain way to Honesdale with an empty boat when he met with a misfortune which will wipe out almost if not all of the profits of his summer's work. He had just passed over the small aqueduct which spans the Lackawaxen River when Train #4 came roaring along at the rate of 40 mph. His team was not used to locomotives, and when they saw the iron steed approaching emitting loud and angry volumes of steam from the cylinders, they began to back up and gave evidence of their desire to escape from the scene.

"Seated on one of the horses was a small boy, the driver of the team. He attempted to keep the horses' heads in line with the towpath, but without avail, and in endeavoring to manage them he was thrown and landed in the canal. In falling, he grabbed the tow line, and the horses in their fright, under the influence of his weight, backed into the canal. When the captain of the boat . . . saw his driver fall into the canal and knowing that he could not swim, he steered the stern of the boat to the towpath, jumped off and ran forward and saved the boy from drowning. In the meantime, the horses were struggling in the water, and the boatmen would have been able to save them had not the boat, with the momentum which it was under bore

DL 4 Courtesy ELLENVILLE PUBLIC LIBRARY

The Dickie Dash awaiting its turn to cross the Delaware Aqueduct on a hot August afternoon. Aboard the snubbed boat are Bernie Eidle of Sparrowbush and John D. Miller of Ellenville. If the boat were running light the awning would have to be removed before passing under bridges.

down upon them and buried them beneath the surface. He saved the life of the boy, but he lost a team of horses valued at over $300."

There was some complaining from the raftsmen at first due to the seemingly narrow 124 foot clearance between the aqueduct piers. The steersmen's skill was challenged in guiding their rafts between the piers, particularly the *doubles,* [12] without touching. After a while they acquired considerable skill in getting through without scraping their timbers. It became a mark of great distinction to get through without, as the raftsmen called it, *scratching.*

In April of 1849 Roebling commenced the High Falls Aqueduct and in June work was started on the Neversink Aqueduct. The High Falls crossing called for the replacing of a twin arch stone bridge while a wood and stone bridge was eliminated at the Neversink River site.

Even after the building of the aqueducts, delays were still being experienced, but this time because of low water due to droughts, as witness this release in the Port Jervis *Gazette* of September 15, 1894. "The many loaded boats that have been so long tied-up between Honesdale and the Twelve Mile Level were let loose on Saturday morning and are now on their way to the Hudson River".

DL 5 Collection of JIM SHAUGHNESSY

A light boat waiting to enter Lock #71 on its three lock approach to the Delaware Aqueduct from the New York side. Lock #72 may be seen in the distance. Note the deck hand at the bilge pump and the bowler-topped dandy at the tiller who is, without a doubt, a son of Erin.

It seems that extremes of adversity were commonplace on the Delaware and Hudson. High waters in 1862 made a sham of a statement released by the *American Railroad Journal* in their July 7, 1849 issue. It stated that "in the construction of the Delaware and Lackawaxen Aqueducts nothing had been spared to insure the safety of the foundations or to furnish efficient icebreakers to guard the piers against heavy flood." [13]

"In June 1862," recorded John Johnston, "by reasons of heavy rains along the sources of the Lackawaxen River, that stream attained a height never before known to the memories of man. The Delaware . . . was but slightly affected by the rains and so great was the

DL 7 Collection of M. B. WAKEFIELD

The two-span Lackawaxen Aqueduct just before the crossing of the Delaware. Out of the picture to the left are the three Ridgeway Locks marking the start of the Lackawaxen Section.

DL 6 Collection of JIM SHAUGHNESSY

Looking down the Delaware River from in front of the Delaware House shortly after the closing of the canal. Just to the right, in back of the young boy, was once located Old Lock #1 of the Lackawaxen section. From this point the rope ferry carried the horses across to the towpath which ran down the opposite bank to the guard lock at the left hand end of the aqueduct.

DL 8 Collection of JIM SHAUGHNESSY

The village of Lackawaxen from the New York side of the Delaware. Above Otter Eddy, shown at the left, is located the Erie Railroad Lackawaxen depot.

DL 9 Collection of JIM SHAUGHNESSY

The Delaware Aqueduct looking upstream over the village of Lackawaxen just after canal abandonment. The Erie Railroad canal bridge was once located just beyond the branch line to Hawley. The triple tiered old Delaware House shows at the confluence of the Lackawaxen River. Located across the Lackawaxen River from the Delaware House is the home of Zane Grey where "Riders of the Purple Sage" and many of his other books were written.

force of water issuing from the Lackawaxen that it rushed across the Delaware and flooded up the opposite shore. In the mad rush of water the pier of the aqueduct spanning the Lackawaxen was undermined and thrown down bodily with but a single crack, the use of the aqueduct destroyed and the entire business of the canal suspended for the time." [14] DL 7, DL 10 & DL 11

With the collapsing of the center pier a large hole about 20 feet across and 10 feet deep immediately filled with the rushing water. Mr. Lord, the chief engineer, was quickly on the scene and immediately erected a cofferdam around the hole. A steam pump was put to work and the water lowered about an inch. Not to be daunted by a badly leaking cofferdam, more pumps were added until six were in operation. The water level went down three inches. Mr. Lord continued these futile operations about nine days more, involving the services of 100 men and holding up hundreds of coal-laden canal boats and their clamorous captains.

At this juncture, John Johnston records the appearance on the scene of a Mr. Sykes, "a principle of a man of the company whose sober and more extensive judgement at once discovered the source of trouble. He suggested that the cavity be filled by throwing in stones plentifully mingled with good cement mortar. It was adopted. One thousand barrels of cement were secured and in six hours after the cement was hard the cavity was filled and the best possible foundation prepared. It is safe to estimate that the errors of a drunken man . . . cost the company not less than $300,000." [15]

DL 10 Collection of JIM SHAUGHNESSY

The Lackawaxen Aqueduct with the Hawley Branch of the Erie in the background. Showing to the right in the streambed is the upstream end of the center abutment. The loss of the abutment at this point in the June 1862 flood closed the canal for over a week. The close proximity of the railroad here and further on around the bend at the crossing of the Erie main line resulted in the loss of many tow horses in the canal when startled by locomotive whistles.

The Delaware crossing stood up well down through the years of flood and drought with the last maintenence work by the Delaware and Hudson Company being completed in 1898.

"Just after navigation closed last fall some three score men, mostly carpenters, commenced tearing out the old planking and timbers and replacing

DL 11 M. B. WAKEFIELD

The west abutment-anchorage of the Lackawaxen Aqueduct still stands today (1965). The middle and the east abutments were demolished for their cut stone content.

DL 12 M. B. WAKEFIELD

The rusting eyebars still emerge from the ruins of the east abutment of the Lackawaxen Aqueduct.

them with new. As it is about seven hundred feet long and required half a million feet of lumber some idea of the magnitude of the work may be imagined." [16]

DL 13 Photo by JIM SHAUGHNESSY

The Pennsylvania upstream anchorage of the Delaware Aqueduct. The eyebars, shown left, secure the splayed-out cable.

The Delaware Aqueduct looking from the Pennsylvania side shortly after it was converted to wagon traffic. The hill in the background was the scene of the bloody Battle of Minisink on July 22, 1779.

DL 14

Collection of JIM SHAUGHNESSY

John August Roebling built exceedingly well when it is realized that the bridge is still being used as an auto toll bridge 117 years later. DL 14 & DL 15

The engineering genius of Roebling had its final flowering when on May 24, 1883, the Brooklyn Bridge was dedicated in New York City.

DL 15 Photo by JIM SHAUGHNESSY

Roebling's Delaware crossing, today shorn of its wooden trunk sides, is captured here by skilled lensman Jim Shaughnessy as a still functioning toll bridge (1965). Evidence of heavy ice floe concentration shows in the ground-up appearance of the abutments upstream surface.

The Delaware Section 6

The Delaware and Hudson Canal through the valley of the Delaware River was in many ways the most spectacular section of the canal. Because of the convolutions of the river and the high bordering mountains, festooned with little feeder streams, nine guard gates and locks were built to aid in controlling possible breaks and washouts.

Four aqueducts were built over streams coursing down to the Delaware: the Beaver Brook, Halfway Brook at Barryville, Big Warwarshaw Brook near Pond Eddy, and the Mongaup River.

Driving down the Delaware from the Roebling Suspension Bridge the first aqueduct site encountered is the Beaver Brook. D 1. The remains of this aqueduct may be seen on the right just beyond a small picnic area. Route 97 passes through Beaver Brook Basin and out along the Delaware River, the road being built directly on the canal right-of-way the entire distance into Barryville.

A post office was established here in 1830 or '31, having been named after William T. Barry, the acting Postmaster General during the first term of General Jackson's presidency. By 1872 it contained two hotels, two churches, four stores, three blacksmith shops, one wagon shop, a drydock for building and repairing canalboats, a grist mill, a stone quarry, a public school, forty-seven dwellings and 259 inhabitants. [1]

The drydock just east of the Halfway Brook was built by John W. Johnston [2] "for N. B. Johnston in the fall, winter and spring of 1850 and '51. After completion it was worked by George A. Johnston and secondly by Horace Tinsman, son-in-law of N. B. Johnston." The boatyard was under the supervision of "Edward Child, who came to Barryville from Wurtsboro in the early spring of 1837 and commenced the building and repair of canalboats. The boats being then small, he drew them from the water on foot-square timbers by means of capstans worked by levers." [3] Later the boatyards were sold "to Charles Frace who moved here from Honesdale and conducted the building and repair of

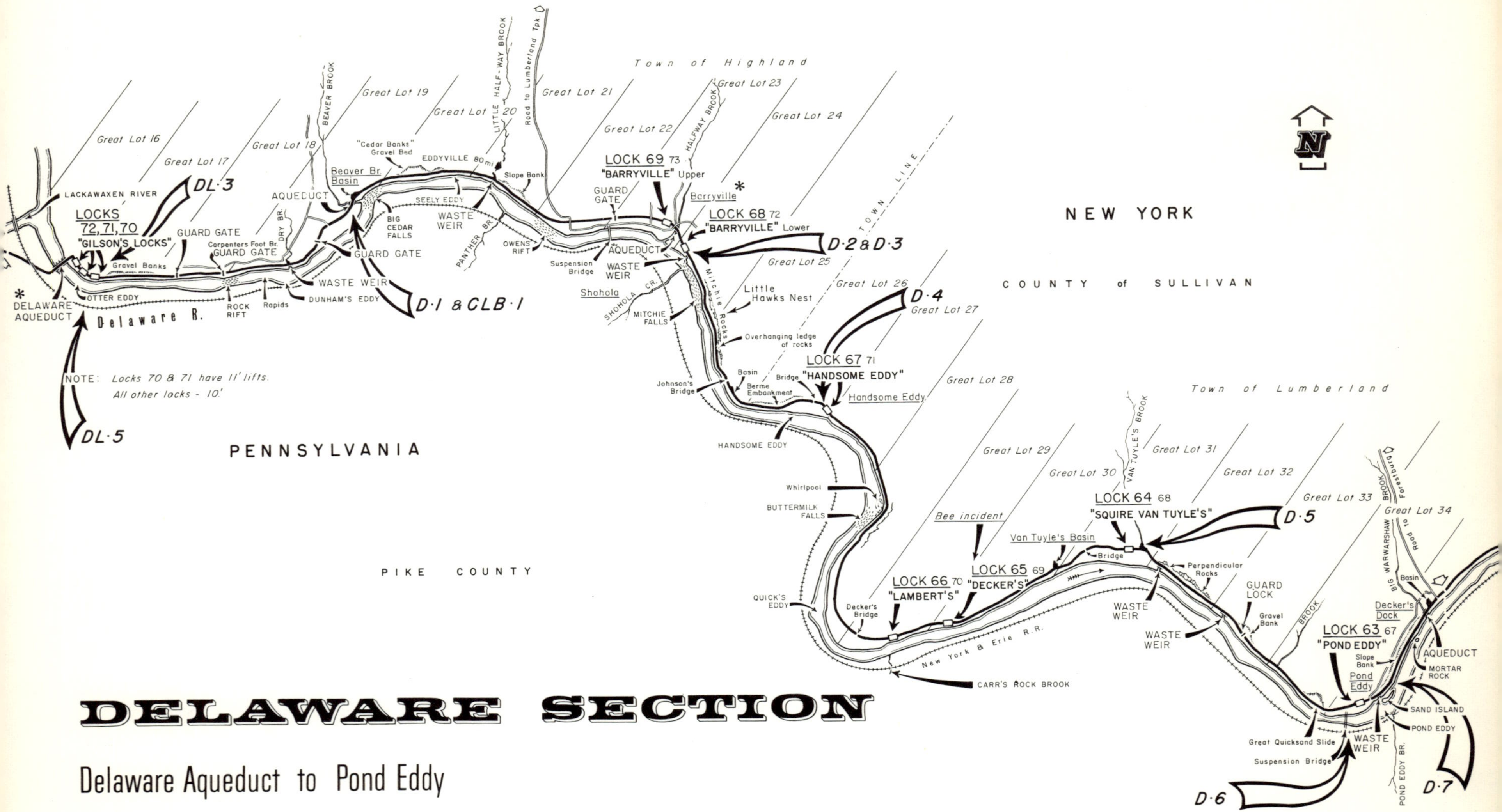

DELAWARE SECTION

Delaware Aqueduct to Pond Eddy

Pond Eddy to Port Jervis

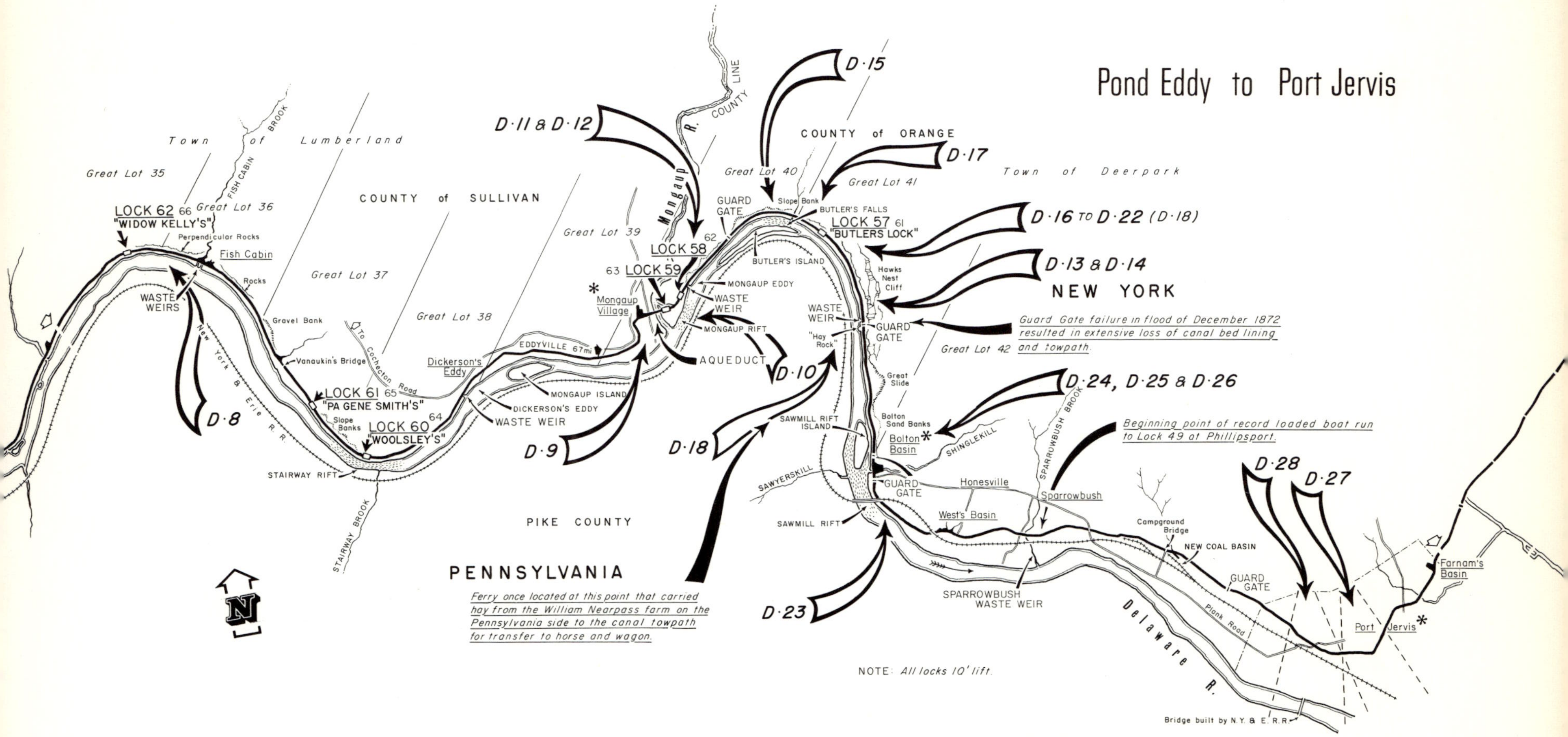

D 1 M. B. WAKEFIELD

The berme side of the Beaver Brook Aqueduct showing Route 97 in the background. To the right, out of the picture, is a small parking area.

canalboats from 1868 until the abandonment of the canal." 4

In 1872 Barryville was connected to Shohola, across the Delaware in Pennsylvania, by a suspension bridge of two spans, 500 feet in length, 10 feet in width and elevated 25 feet above the water. The towers on which it rested were 45 feet in height, and were contracted for and built by Chauncey Thomas of Shohola at a cost of about $4,000. This was the third bridge at this point. The first, built in 1856, blew down three years later. It was rebuilt and shortly thereafter collapsed with a broken cable. Finally, the year 1866 saw the completion of this, the last suspension bridge at Barryville.

"Passing from the bridge (aqueduct) about four rods down the canal, Lock #68 (#72, Barryville, lower) is reached. Nearly opposite the lock and high up on the hill, towards the east, is the lock house, first occupied and locktended by Zibe Ridgway; next by Robert Johnston; then by Iser Rake for a period of forty years, and finally by Jacob Beck." 5

Lock #69 (#73, Barryville, upper) was tended by Michael Libla in 1872 - 73. Other canal employees at that time were section foremen John Barnes and John E. Lounsbery and the well-liked canal superintendent, James H. Quick.

Lock #68 is still very much intact, although badly cluttered with junk and debris at the drop gate end. D 2. An original canal-side store (#6 on BARRYVILLE map) still stands, while just below it the original fencing indicates where a D. and H. Canal Company building once stood. D 3. Opposite the basin below the lock, the stone foundations of the barn and stables may still be seen (1965).

Downstream from Barryville the canal is hemmed in between the plunging Mitchie Falls and the towering Mitchie Rocks. At the lower end of these rock formations, high on the cliff, was once found the famous Hanging Rock. In 1868 the rock was ordered removed. As John Johnston tells it, "The rock was a part of the Midchuck ledge, massive in proportions, about 75 feet high, 50 feet in length, up and down the canal; quite level on the top and about 35 feet wide . . . its projection over and across the canal was 28 feet, reaching within four feet of the towpath of the enlarged canal. It contained 10,000 cubic yards by actual computation. For a few years prior to its demolition, fears arose with the canal authorities that it was dangerous; that it was liable to break from

M. B. WAKEFIELD

Barryville Lock #68 showing guard rails of Route 97 in the background. A charred member of the wood lining shows at the right.

Barryville

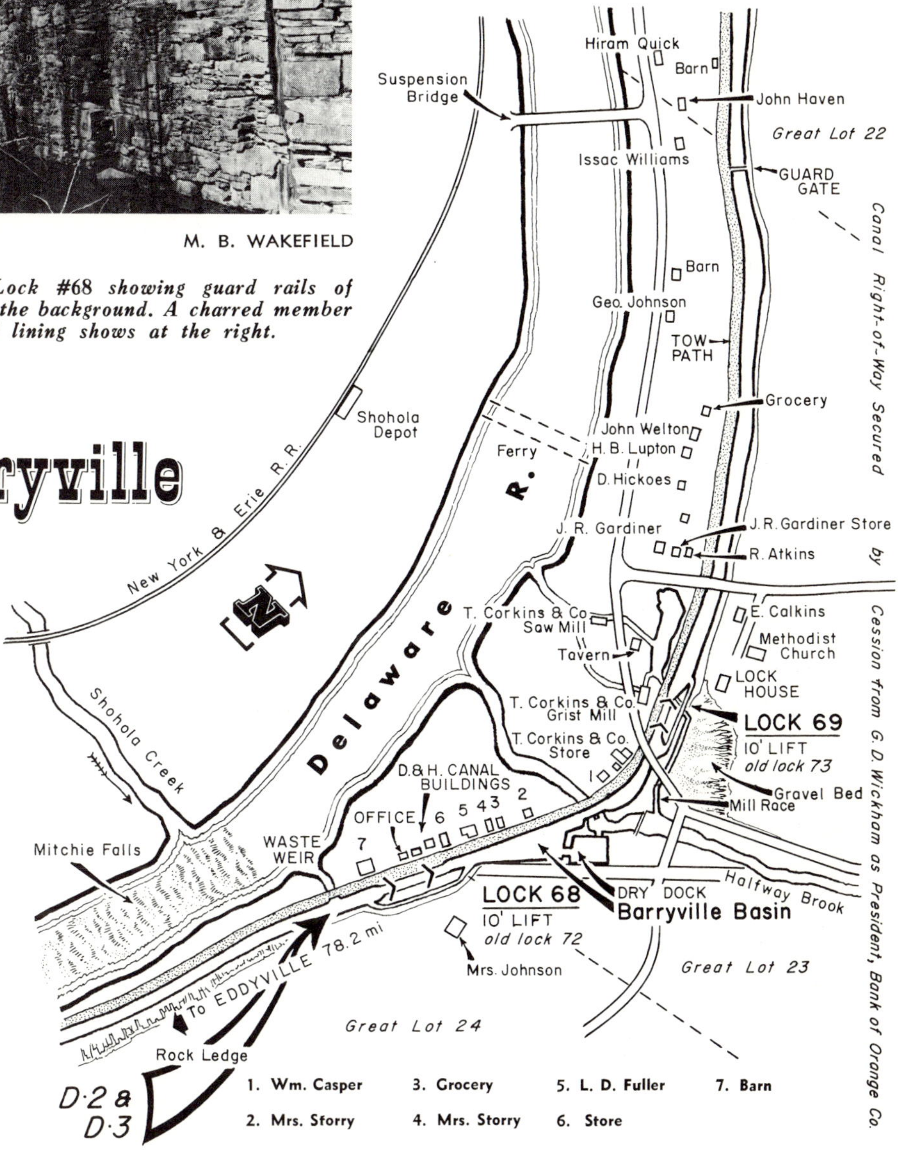

its holdings and fall . . . In the autumn of 1868, after the canal navigation had closed, it was ordered that 6 or 8 drill holes of large size and ample depth to be (sic.) inserted in different parts of the top, believing that when charged with powder and exploded, the jar would loosen and throw down the rock . . . Considerable fragments were thrown out, but the rock not disturbed.

"Back of the rock was an ample seam and that was next loaded with 28 kegs of powder (700 pounds) and buried under eleven feet of tamping, thoroughly packed. When the explosion occurred a vast quantity of stones were thrown, some beyond the Pennsylvania shore of the river, and the long and well known Hanging Rock fell to rise no more." [6]

D 3 M. B. WAKEFIELD

The towpath at Barryville Lock #68, right side of picture, was lined with D. and H. Canal structures. The building in the background was a canalside store while the old iron fence marks the site of the D. and H. Canal Co. office in Barryville.

D 2 M. B. WAKEFIELD

Today Lock #68 at Barryville, in addition to containing a water logged boat, is the recipient of a considerable quantity of garbage.

Just below the Mitchie Rocks the canal curved into Handsome Eddy Lock #67. D 4. The placid eddy was a scene of chaos in May of 1832. "About the 8th of May, rain commenced falling, and fell with violence for nearly three days and nights and the Delaware became a raging flood unequaled in the memory of the oldest inhabitants. For one whole day the surface of the river was literally covered with rafts and lumber of all descriptions, and of all the valuable rafts anchored at Handsome Eddy but a single one was saved; that of which David Quick had charge; not by superior management but owing to a favorable position. That high water has been ever since known to the older class of people along the Delaware as the

"May Flood" and unquestionably the highest ever known until the flood of 1869." [7] Mr. Johnston also remembered great floods in 1834, '37, '41, '52, '69 and '95.

For the next mile or so below Handsome Eddy the river "sets heavy" against the New York side. This extensive sag terminates in what was once known as the Whirlpool just at the edge of Buttermilk Falls, also recorded as Blackmore's Rift.

The Whirlpool "was at the edge of the river, at a slight bend in the stream, confined to about a half acre and whirling at the center seemingly sufficient to engulf whatever floating object is caught by it. The water commenced to circle the outer edge, with a radius of about 60 feet and continued with regularly increasing violence until, at the center, it formed the shape of a tunnel and seemed as though discharged through an opening at the bottom, thus forming a perfect maelstrom." [8]

At times of freshet the area would trap tens of thousands of feet of logs and debris floating downstream. John Johnston has known "50 to 75 thousand feet of good lumber to be secured at this point in a single day by means of

D 4 M. B. WAKEFIELD

The well tended canalbed and towpath at Handsome Eddy. The filled-in crossing below the footbridge was Corwin's Bridge in the 1850's. The barn to the right was used as a stable for Lock #67. This lock, out of the picture to the left, has been completely filled in.

pike poles and hooks." Frequently people fell in the river upstream only to be swept into the vortex of the whirlpool and drowned.

On the berme side of the canal at this spot a large stone quarry was later established. The quarry operators made the whirlpool a deposit for the rubbish and debris from cutting, thereby obliterating all traces of the whirling phenomenon.

Farther downstream, after going around the big bend at Quick's Eddy, Lock #66 is encountered, which was also known to the canalmen as *Lambert's,* followed closely by Lock #65, known more popularly by the name of *Decker's Lock.*

It was here on a hot August day in 1870 that a lively time was had due to the upsetting of two skips of bees. "The bees became so enraged," according to the *Tri-States Union* of August 4, 1870, "that they swarmed upon everything near them, making it impossible to boat for a long time. Teams ran away, boat boys jumped into the canal, captains took to the woods, and for a time all was confusion and excitement in trying to get out of reach of the angry insects, who seemed determined to wreck vengeance upon all who came near them."

Past Squire VanTuyle's Lock #64 (D 5), one half of which has been filled in for a driveway, the highway swings up and over the high shoulder of a mountain that terminates in a towering cliff over the canal.

D 5 M. B. WAKEFIELD

The remains of VanTuyle's Lock, Lock #64, looking toward the drop gate end. Unfortunately the lower gate end has been filled in for a driveway approach to a house. Just below the lock, VanTuyle's Brook fed water into the canal.

Pond Eddy, the next community on the line of the canal, contained one church, one small school and fifteen dwellings in 1872. It wasn't until that same year that a post office was finally established. Pond Eddy did boast a suspension bridge with a span of 521 feet and a width of 12 feet. D 6. It was elevated 26 feet at the

abutments and 31 feet at the piers above low water. The structure was erected by James D. Decker, who was a contractor for the Town of Lumberland during the years 1870 and '71. In 1872, the appointed bridge gate keeper was John Main.

The hamlet was a mecca for canal boatmen including James M. Covert, Cornelius Foley, Jacob S. Pool and John Steel. The D. and H. Canal foreman on post at Pond Eddy in 1872 was George H. Langton and the canal watchman was James VanGorder. Pond Eddy Lock #63 was tended by Cornelius Van-Gorder. D 7

At many points along the canal were places known as deer runways, one of which was located just below Pond Eddy. John Johnston related the story of " . . . Jeremiah Sears of Pond Eddy who had an insatiable thirst for hunting and killing deer. On a certain occasion he was watching a runway at the lower end of Pond Eddy, in expectation of a deer fleeing from the hounds and seeking protection in the water of the river. A canalboat on its downward passage was approaching the runway at the usual slow movement, and the negro steersman was loudly blowing a good 6 feet (sic.) tin horn, a common amusement and practice of boatmen, the large end of the horn resting on the cabin in front, and the small end and mouth piece at the lips of the darky.

D 6 Collection of M. B. WAKEFIELD

The Pond Eddy suspension bridge crossing of the Delaware River. The spectators standing on the canal bridge are no doubt watching the rise of the Delaware which appears to be at freshet stage. The bridge was built in the years 1870 - '71.

D 7 Courtesy of ELLENVILLE PUBLIC LIBRARY

Lock #63 at Pond Eddy shortly after the closing of the canal and the draining off of water. The old Taylor house is shown on the right which in the 1854 maps was shown as the lock tender's house. The structure with the double-decked porch on the left was a tavern and inn.

"Sears, fearing the sound of the horn might divert the deer from the runway, walked up the towpath until he met the boat and kindly asked the blower to desist until further down the canal stating his reasons for the request, but according to the habits of boatmen the request was met by an invitation to Sears to go to Hell with his deer etc., after which the horn was sounded with increased volume. Thereupon Sears, without another word or the lapse of a moment, sighted his rifle and severed the horn within less than a foot of the negro's mouth; with the exclamation, 'Now there, God damn you, toot again, will you!' The darky stood bewildered with his tongue and horn both effectually silenced."

Route 97 passes through Lock #62, *Widow Kelly's,* one wall of which is still standing. The river side of the road for quite some distance is retained by the original stone work thrown up by the D. and H. construction crews in 1827. D 8

D 8 M. B. WAKEFIELD

Lock #62, known to DuBois Weber as the Widow Kelly's Lock, was located just to right of center (ARROW) still showing a portion of the inside lock wall. The outer towpath stone work is still in place supporting and slightly raising Route 97 in the lock area.

D 9 Collection of GERALD M. BEST

The Mongaup Feeder Dam showing the old suspension bridge crossing. The water fed into the canal to the right from a point between the dam and the bridge.

Downstream the next center of canal activity was Mongaup Village where the canal left Sullivan County while crossing the Mongaup River into Orange County. In 1872 it contained one store, a telegraph office, a school house, a boatyard for building and repairing canalboats, ten dwellings and fifty-four inhabitants.

Its primary reason for existence was the canal and the large aqueduct for the crossing of the Mongaup. Two locks, #58 and #59, were built here to raise boats up for the Mongaup crossing. A feeder dam was also erected across the Mongaup for diverting water into the canal. D 9

The widespread floods of December 1878 created some concern at Mongaup as noted in this December 12 story in the Port Jervis *Evening Gazette*. "The Mongaup River ran over its banks at the upper part of Mr. West's farm and tore down through the land, into the canal, tearing out the berme wall. The Mongaup Aqueduct is somewhat impaired by the washing away of fifteen feet of the upper part of the pier, but the superstructure is all right. At Pond Eddy forty feet of the towpath river wall was washed out, leaving but five feet of towpath intact."

D 10 Collection of GERALD M. BEST

Lock #59 at Mongaup raised the boats up 10 feet for the crossing of the Mongaup Aqueduct. The Mongaup feeder enters from the right, passing under the foot bridge into the large basin.

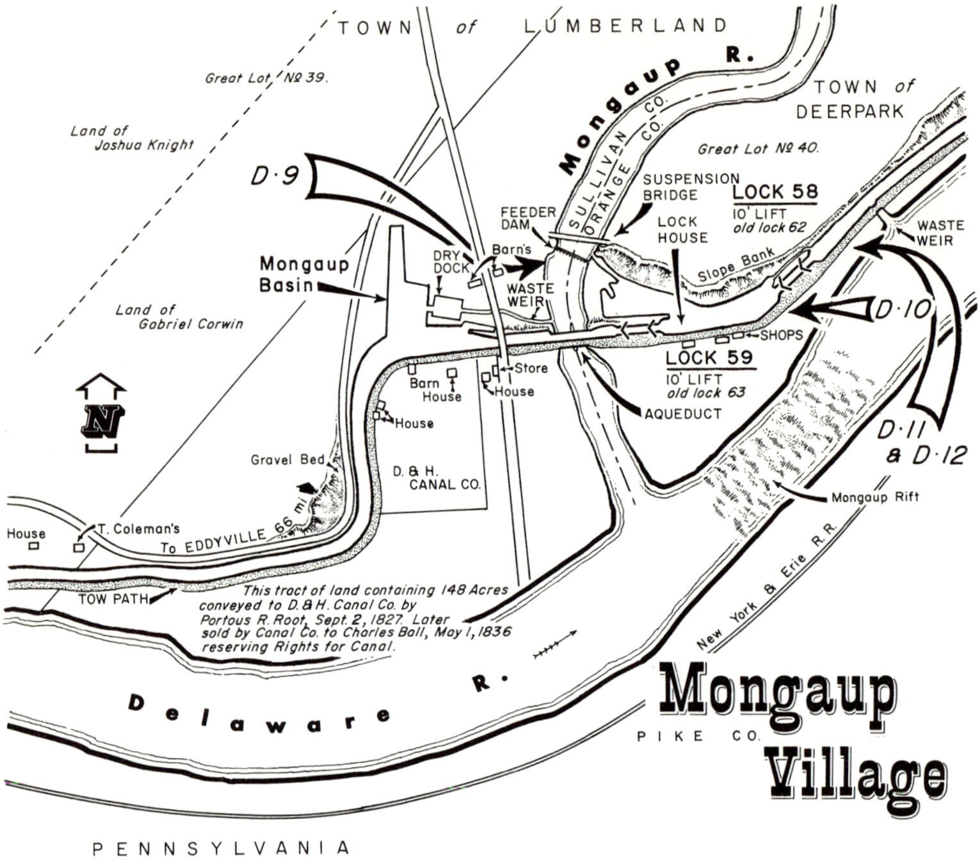

During the early 70's the Mongaup Locks were manned by W. H. Couse and Jackson Johnson. Canal foreman was Daniel S. Coleman, the superintending foreman was H. K. Corwin, and the canal watchman John Johnson. Mongaup had its resident boatmen and boatbuilders too. Boatmen Asel Adams and Nathan Adams listed Mongaup as their home while Martin May and George West were busily engaged in boat building. West launched two to three new boats a year at his dock.

It is difficult today to imagine the bustle and activity of this canal town. Route 97 crosses the Mongaup just below the canal crossing. Virtually all traces of the aqueduct have vanished. The fill for the highway bridge approach has covered Lock #59 and part of the feeder area. D 10. The stone work in Lock #58, a short walk downstream from the highway, is as straight and true as the day it was built and entirely free of debris. D 11 and D 12

DuBois Weber relates an incident that occurred just below Lock 58. "During a bad lightning storm my father would always say to me to get up on a horse next to the canal and you won't get hit. One day between Mongaup and Butler's Lock I see lightning hit the top of a telegraph pole just ahead of me and pieces fly down into the river and canal. I was riding on the lead

D 11 M. B. WAKEFIELD

The remarkably well-preserved stone work at the lower gate end of Lock #58. The shadowed vertical recess in the wall that held the gate shows just to the right of the tree in the lock.

mule. My father often said bad lightning storms always follow the rivers and canal." [9]

The lower part of the Delaware became the scene of conflict and embarassment to the canal operators in 1847 and '48. As the construction crews of the New York, Lake Erie & Western Railway moved further up the opposite side of the river, so too did the frequency of altercations with the canal boatmen. The men hired by the railroad contractors appear to have been a most desperate and lawless gang. They fought among themselves, resulting many times in loss of life.

Having firearms in their possession they took sport in directing random scatterfire at canal boats passing by on the other side. The boatmen

D 12 Collection of GERALD M. BEST

Mongaup Lock #58 shown with a considerable amount of water running over its upper gate sill. This quantity of water is no doubt coming from the Mongaup feeder and being permitted to flow directly down the canal. Wagon tracks on the towpath would indicate the canal had ceased operations.

D 13 Collection of GERALD M. BEST from WILL ALLERTON GLASS PLATE, GUS KRAUSS Collection

Looking from the Hawk's Nest Cliff showing Butler's Lock #57 in the distance. Directly over the train on the Erie tracks may be seen the lock tender's house for Butler's Lock. This lock was in part located at this spot because of the huge rock outcroppings along the shoreline. D 22

who were obliged to run the gauntlet of pistol balls and other missiles were becoming increasingly annoyed. A feud rapidly developed between the two parties which blossomed into a serious conflict near Mongaup. The battle was a desperate one with many of the combatants on both sides receiving a terrible beating with a number dying of their injuries. A force of police from New York City was brought to the scene of conflict as well as the sheriffs of Sullivan and Pike Counties. The patrol continued until the middle of November.

As construction continued, the blasting crews hit upon an insidious plan. They would locate their powder charges and time them to coincide with the passing of canal boats on the other side. The resulting blast would shower down rock and debris on canalboat and towpath alike. The continued blasting and harassment had a demoralizing effect on the boatmen. Parents concerned over the safety of their children forbade them to hire out as drivers.

Tensions mounted until one day, after a canal-directed blast, the railroad crews were astonished to see swarming out of the canalboats across the river seemingly hundreds of irate, club-wielding boatmen. They stormed down the bank, waded the shallow river and charged up the Pennsylvania side to do

D 14 M. B. WAKEFIELD

Today all that marked the great waterway's course at the base of the Hawk's Nest has vanished completely when viewed from scenic Route 97. Only by clambering along the shoreline is it possible to detect the few tangible ruins that have resisted six decades of floods, floes and frost.

battle. According to John A. McGinnis, "A number of heads were busted but it pretty much settled the whole affair."

Lock #57 marks the starting point of the 12-mile level which passes through Port Jervis and terminates at the Neversink Locks. Rarely if ever was it referred to as Lock #57, perhaps only on official reports, since it was known far and wide as *Butler's Lock,* the lock jammed in at the foot of the Hawk's Nest Cliff.

Butler's Lock, one of the most isolated on the canal, was named after General Butler, a white Tory associate of the notorious Chief Joseph Brandt of Battle of Minisink fame. The canal waters and towpath were retained at the foot of the cliff by a towering forty foot stone wall built up from the bed of the Delaware River. D 13

The top of the stone retaining wall was capped off with smooth flat stones. John A. McGinnis recalls, "You take at Butler's Lock; it was at least thirty or forty feet from the towpath down to the river, and they had a wall about three foot high that was like a guard rail, and of course that was a dandy

D 15 M. B. WAKEFIELD

The remains of the lock tender's house for Butler's Lock is located about a half mile upstream from the lock. The canal berme wall shows in the foreground.

place to walk in your bare feet; why it was just like on top of the table, but if you stubbed your toe you were liable to go down. The mules would be walking along there and as soon as my father would see us on the wall he would holler, 'get off that wall,' and the old mule would come over and walk along the wall until you got on her back, and then she would go on the inside of the towpath again." D 16

Although Butler's Lock and the flat-topped canal wall conjured up pleasant memories for Mr. Weber, it was a scene of tragedy for a small canal boy who "while attending his horses during the passage of his boat through the lock at Butler's, became entangled in the towline. The horses becoming frightened, ran, dragging the unfortunate boy along the towpath, mutilating him in a horrible manner. The horses were stopped after running about

D 16 Collection of GERALD M. BEST

Looking up the canal toward Butler's Lock at the base of the Hawk's Nest showing the massive dry wall thrown up to impound the canal water. Towpath boys and girls always found the top of the wall fun to walk along.

D 17 M. B. WAKEFIELD

The canal along the Delaware between the lock tender's house and Butler's Lock is devoid of its towpath, having been reduced to a small stone wall.

a mile, and the boy was found to be quite dead. His name was John McGrain and was fourteen years of age. The back of his head was, on examination, found to be smashed in." 10

In the spring of 1965 the author had occasion to visit the site of Butler's Lock. The towpath and canal continues but a short distance downstream from the remains of the lock tender's house. D 15. In picture D 17, about halfway between the tender's house and the lock, the left side of the picture reveals the towpath reduced to a mound of stone rubble.

From this point on the going is extremely difficult because of the necessity of climbing up and down rock outcroppings that at one time were covered by the canal and towpath. After what seems an interminable length of time the remains of Butler's Lock, #57 appear. Only the berme side lock wall and overflow weir remains. A small segment of the approach towpath lock wall at the lower gate end still stands. D 19 and D 20. Butler's Lock was built atop large rock outcroppings which have reappeared as noted in the before and after sequence of pictures D 21 and D 22

Below Butler's Lock was located a rock outcropping known as *Hay Rock*. It extended out into the river beyond the towpath and was known to the boatmen as the halfway point between Butler's Lock and Bolton Basin. D 18

Hay Rock received its name, according to William Darragh of Sparrowbush, because a ferry operated across the river at the rock for the carrying of hay from the extensive Pennsylvania farm holdings of William Nearpass. Today these flat farmlands have reverted to a forest of pine trees. Interestingly enough,

D 18 M. B. WAKEFIELD

The remains of the landmark 'Hay Rock' below the towering Hawk's Nest. The rock loomed much higher during the canal years permitting the canal to pass between it and the cliff wall.

D 19 Collection of M. B. WAKEFIELD

Butler's Lock showing the lower gate and outlet for the waste weir run-around at right.

D 20 M. B. WAKEFIELD

Today, from a slightly closer vantage point, only the right-hand lock wall and waste weir remain. A small segment of the towpath side of the entrance wall may be seen on the left.

D 21

Collection of M. B. WAKEFIELD

A light boat waiting just below Butler's Lock with a parasol-equipped party of excursionists on board. Note the loaded boat moving down the canal in the distance.

D 22

M. B. WAKEFIELD

Looking downstream at the rock upon which Butler's Lock and towpath was constructed. The Hawk's Nest Drive, Rt. 97 shows on the side of the cliff.

the hay was not picked up by canal boats but was hauled away by horse and wagon. Although officials took a dim view of wagons using the towpath, the practice nevertheless persisted.

Just adjacent to Hay Rock the canal company installed a guard gate. It was this guard gate that went out in the previously mentioned flood of December 1872, resulting in the washing out of the lining of the canal and towpath. The guard gate at Bolton Basin held, preventing further destruction.

These guard gates, also known as stop locks to the canalmen, rested on the bottom of the canal when not in use, to be cranked up only in an emergency. The gates fell in both directions so that the canal on both sides would be protected as the need arose. When a break did occur the section involved could then be isolated, drained off and repaired. Otherwise, long sections of the canal would be in jeopardy, such as the 17-mile Summit Level and the shorter 12-mile level.

In the event of a break an emergency scow, always on stand-by duty, would be called out. This boat had absolute right-of-way over all other boats on the canal, and carried pilings, planking, wheelbarrows, shovels and all the paraphernalia necessary for stemming the tide at a breach. After a temporary wall was erected across the break, dirt was tamped in on the outside and allowed to settle. N 5

At the lower end of the Hawk's Nest Cliff, just above the Bolton Basin area, a large landslide took place on April 11, 1881. A mass of rocks and earth covering an area over one hundred yards square crashed down into the canal, completely filling it above the water line. Work was immediately started on the removal of the debris by a considerable force of men under the direction of Mr. Corwin, the canal boss. A mile of water on the 12-mile level was drained off to facilitate the job.

The Erie Railroad crosses the Delaware River and the canal at the downstream side of Bolton Basin. The close proximity of the railroad to the canal, plus an errant load of walnut logs and a particular type of bridge construction, all added up to a calamity the likes of which

D 23 Collection of GERALD M. BEST

New York, Lake Erie and Western Railway bridge #2 over the canal prior to 1881. The tracks are of the original six foot gauge width. This bridge was later demolished by a flat car load of logs. A house at the lower end of Bolton Basin shows just over the edge of the bridge.

was never recorded before or since in the annals of the Delaware and Hudson Canal Company.

A contemporary newspaper of October 1882, reported, "This morning at 5:50 o'clock an accident involving the loss of a large amount of property took place at bridge number two on the Erie Railroad, three and one-half miles west of this station (Port Jervis). An extra, number 36 with Will Strader conductor, broke in two, three-quarters of a mile west of the bridge where there is a heavy down grade . . . the two sections coming together on the bridge.

"Two flat cars loaded with large walnut logs, broke from the fastenings that held them and the logs projected over the car, stripped the railing along the bridge until the cars reached the portion crossing the Delaware and Hudson Canal. At this point there was a superstructure of iron girders and braces that supported the bridge. D 23. The huge logs, crashing into this network of iron, tore it from its fastenings; the weakened bridge gave way, and fifteen cars were hurled into the canal. Telegraph poles, both on the Erie and canal, were torn down and for a short time communication by this means was destroyed.

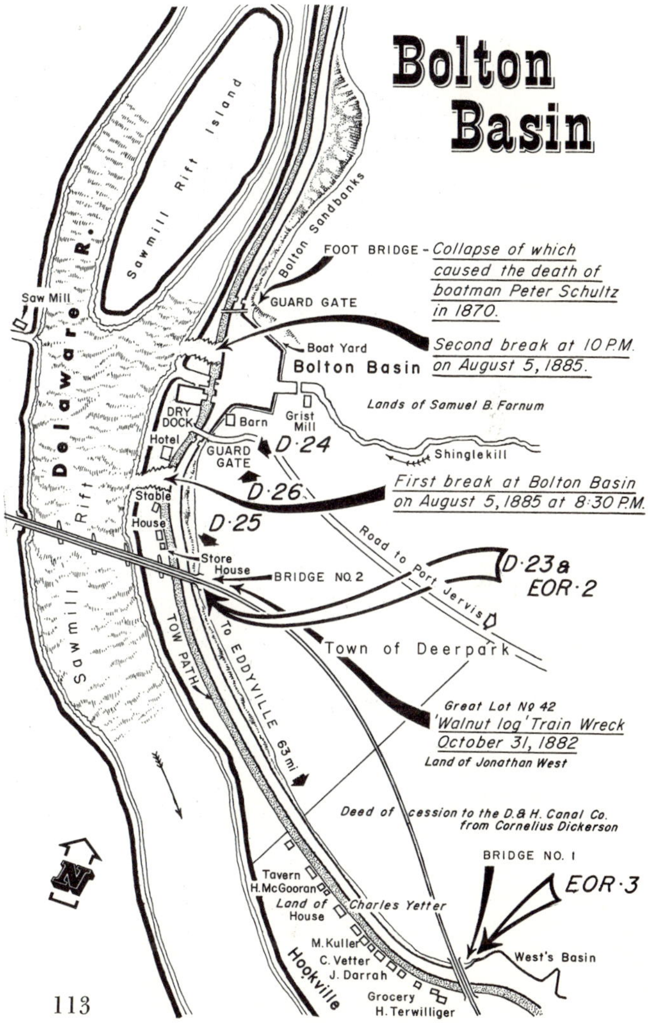

"The train was composed of seventy-one cars, all heavily loaded, drawn by engine 332. Linn Havens engineer. The cars that went down with the bridge were ten loaded with wheat and oats, two of flour, two of walnut logs and one of oil cake.

"They began drawing off the water on the canal about noon. Superintendent Murphy and M. M. VanVechten with a large force of men are working earnestly to clear away the debris and cargoes." [11]

The general scattering of debris and the novelty of a freight train in the canal attracted hundreds to the scene from Port Jervis. In spite of the efforts of officers and employees, every train that ran out to the wreck and back was loaded with sightseers, including women and girls. Most of the water was drawn from the canal to facilitate the removal of bags of flour and the twisted iron of the collapsed bridge.

Bolton Basin was named after the second president of the Delaware and Hudson Canal Co., John Bolton, and soon became a favorite layover spot for the boatmen. They knew they would have good accommodations if they could get to Bolton Basin before the locks closed.

However placid things seemed generally, calamities big and small seem to have had a way of happening at Bolton Basin. The *Tri States Union* of May 19, 1870, reported the tragic death of a boatman named Peter Schultz. "It appears that a boat was passing under a foot bridge at that place (Bolton Basin), when the tiller or the rudder came in contact with the bridge, which is said to have been a frail structure, pulling it down upon the man and crushing him to death. Schultz was a married man and lived in Honesdale, Pa."

D 24 Collection of J. A. O'BOYLE, from original print.

Bolton Basin break number one occurred between Mrs. Simpson's Hotel and the stable at left. Mrs. Simpson is shown standing just to the right of the wagon between the two men.

"The northeast storm which began in this section Sunday afternoon continuing yesterday, became almost a tornado before night. The rain fell in blinding sheets, accompanied by heavy winds and in the evening by thunder and lightning and everything was flooded."

So opened the lead story in the Port Jervis *Tri States Union* of August 6, 1885. This same widespread storm was also the agent that precipitated the now famous Bolton Basin Break. The story goes on to relate, "The greatest damage done was to the Delaware and Hudson Canal. At Bolton Basin the embankment is badly washed out at two points. The first one was just east of the hotel kept by Mrs. Alex Simpson, where for 100 feet the bank is gone out to a depth of 20 feet below the bottom of the canal."

The first break took place at 8:30 Monday evening, August 5, twenty-four hours after the storm started. The precise time was noted by Mrs. Simpson because the roar of the rushing water, coupled with the grinding of the boats and stones, made such a terrifying noise that she knew something terrible had happened. Rushing out on the porch of the hotel, the pandemonium told her the break had occurred very near to the hotel. D 24

D 25 Collection of J. A. O'BOYLE, from original print.

Looking down Bolton Basin past the first break, shown at the right, towards the new Erie bridge built to replace the one demolished in the famous 1881 walnut log train wreck. The two hex signs on the boats to the left evidently worked since the boats remained upright. Note the two little girls on the bow.

The Shinglekill which feeds into the basin is normally a quiet brook noted for its trout fishing. The downpour soon had it a roaring cascade breaking its bank at many points.

The *Tri States Union* went on to explain. "There are several waste weirs opposite the point it (the Shinglekill) enters the canal, all of them designed for just such emergencies, and had they been lifted the water would all have passed out without damage. As it was they soon filled up with brush and driftwood, brought down by the raging Shinglekill, and formed an effectual dam. Then it was that the first break occurred."

William Darragh recalls that all steps were taken to permit the water to escape, but that the canalboats in the basin were swept across the waste weir effectively blocking the passage of water.

There were a large number of boats tied up in the basin for the night, and when the break occurred many were caught in the surge of water sweeping through the basin to the breach. Three of the boats became wedged together

D 26 Collection of J. A. O'BOYLE, from original print

The jumble of boats looking upstream through the basin toward the upper foot bridge and guard gate. The second break occurred just before the second building on the towpath to the left. The footbridge in the background is the second at that location. Collapse of the first resulted in the death of Peter Schultz in 1870.

just west of the bridge over the canal in front of the hotel. This was the grinding together of canal boats that Mrs. Simpson heard over the howling of the storm. D 25. This effectively blocked off the flow of water to the break, backing the swirling water up until about 10:00 p. m. when the second break occurred about one thousand feet to the west of the first break.

"This is even more serious than the first," the report went on to say. "The bank for 200 feet is carried away and with it earth from the bottom to a depth of 20 to 25 feet. At this point a boat loaded with coal was broken from its fastenings and carried through the gap for 200 feet and nearly out to the waterline of the Delaware." That boat was owned by Lambert Artman and was a total loss. There were nineteen other boats in the basin, twelve being loaded and eight empty. Six of the loaded boats were so badly damaged as to be worthless. Two of the light boats were badly sprung and virtually beyond repair. Before the breaks occurred the boatmen and their families left their boats as the waters started inching towards the top of the towpath.

The next day the boats lay twisted around in all conceivable directions, with the waters of the Shinglekill still rushing through the great gaps in the towpath. The canallers sat in disconsolate groups on their boats or lounged upon the canal banks mourning their bad luck. Mrs. Simpson wasn't brooding over the disaster as she was much too busy waiting on callers and sightseers at the hotel who came to view the chaos. D 26

Viewing the half mile of bottomless canal, the canal boss, Mr. H. K. Corwin, and Mr. A. M. Montgomery, the canal carpenter, agreed that a worse break had never occurred along the line of the D. and H. Canal.

In Port Jervis on June 6, 1862, failure to raise a guard gate resulted in a flood with losses of upwards of $20,000 to the business and residential community.

"The unprecedented rains of Monday, Tuesday and Wednesday filled the canal to overflowing, causing a break a short distance above this village, and inundating the whole lower part of town. The break occurred shortly after seven o'clock on Wednesday evening, and immediately the streets were flooded. Basements and cellars filled so rapidly as to preclude the possibility of saving anything. From Wells' store on Pike Street to the depot was a swiftly running stream tearing up sidewalks, damaging streets, and undermining foundations of buildings. The whole eastern side of the street is seriously damaged. The great body of water was finally thrown through Ball and Hammond Streets which became impassable." **12**

By eleven o'clock the guard gate was raised allowing the waters in the village to subside, revealing the break in the towpath to be in excess of five hundred feet in length.

"There were several narrow escapes from drowning during the flood. Mr. J. G. Brooks, in attempting to save some articles in the cellar of Everitt & Noony's store, fell from the box on which he was standing, and plunged into the water which was within a few inches of the first floor. Johnnie, being somewhat amphibious in his nature, succeeded in dividing his time between air and water until a place of exit presented itself.

"Mason D. Newman, at the store of O. F. Sherman, had a very narrow escape. While passing into the cellar, the steps gave way beneath him, and, but for the timely assistance of Patrick Flannigan, he would have been drowned. At the corner of Sherman's store on Pike Street, several persons were rescued with great difficulty. The force of the current had washed the street away, and left great cavities which in the darkness were not noticed." [13]

On June 27, 1862, chief engineer R. F. Lord announced that the canal would be ready for business by July 1st. The consternation of the boatmen over a twenty-six day boating delay can well be imagined.

Port Jervis did experience seasons when there were few if any delays to boating. In 1887 there was only one delay. "This was on the afternoon of

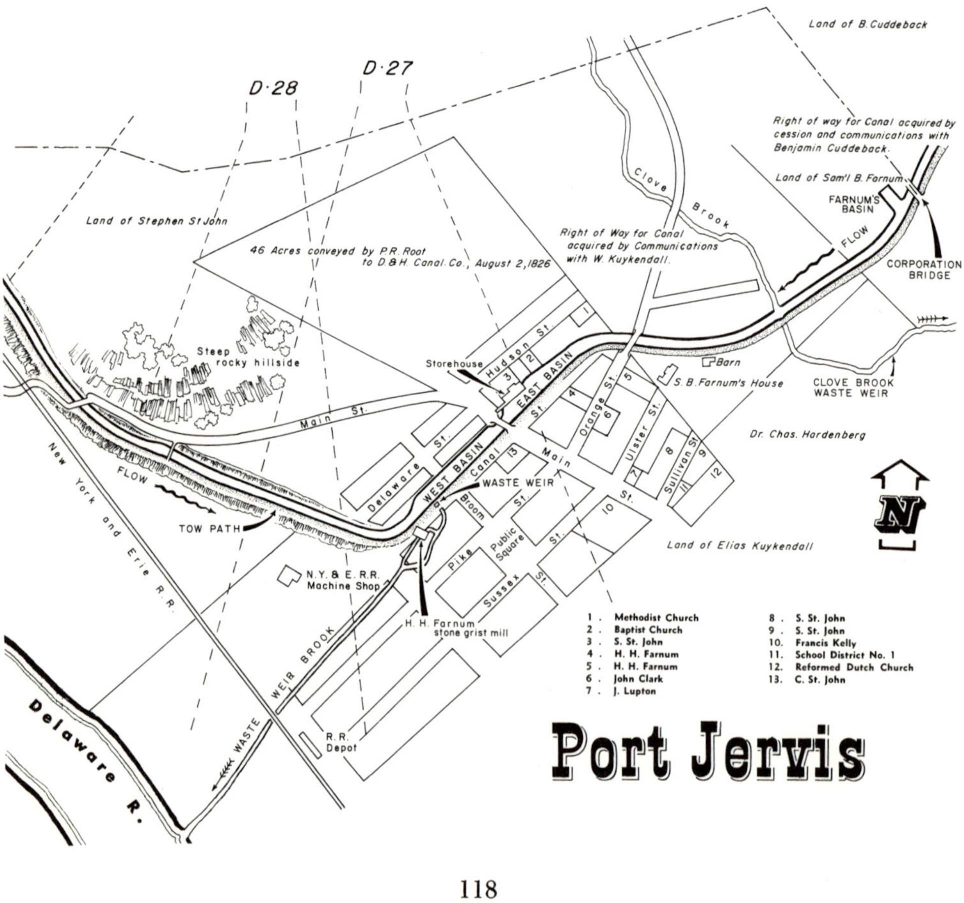

November 18th when Supt. L. O. Rose ordered the water drawn off at Port Jervis so that the people of that village might make a search for the body of Richard Gray, president of the Orange County Building and Loan Association and member of the Port Jervis Board of Health. The water was lowered in about four hours and the search got under way shortly after dark. The old canal presented a weird scene as people searched up and down by the dim lights of lanterns and torches. At 9:30 p. m. the search proved successful when the body was found on the berme side of the canal, opposite the barn of Samuel B. Farnum. Dawn saw the canal bed refilled and navigation resumed." [14]

Many people, particularly strangers, became victims of drowning in the canal. As the list of casualities grew, so did public sentiment. Finally, on April 26, 1890, when the body of David L. Baird was fished out of the canal, Coroner Harding petitioned the Port Jervis Board of Trustees to erect guard rails along the canal. The proposal was found to be impractical and was never carried out.

In 1880 what was at first feared to be a drowning turned out to be a sensational kidnapping that resulted in the arrest of a boatman at Rowlands on the Lackawaxen.

Two nephews of Mr. William Carter were playing by the canal in Port Jervis when one went home for a moment only to return later to find his cousin gone. The *Tri-States Union* of September 21, 1880 reports in part, "Mr. Carter, who is a conductor on the Eastern Division of the Erie Railroad, returned home yesterday and at once was informed of the state of affairs. Mr. Carter immediately began a systematic search. Thinking that possibly the boy had gone away on a loaded boat, he took a carriage and drove to the Neversink Locks where forty or fifty boats were lying . . . At length he came to a boat on which was a family. The woman asked him what he was searching for, and he told her. . . . She gave an accurate description of the boy's dress and stated that he was on William Duffy's boat #1717. . . . He drove immediately back to Port Jervis and learned that the boat 1717 had not yet passed Rowlands on the Lackawaxen Division. He then telegraphed to parties at that place to search the boat and take charge of the boy answering to the name of Abbey Willis. When the boat reached Rowlands Lock, the parties to whom he had telegraphed boarded the boat, entered the cabin, and found the boy crying bitterly." [15]

In Port Jervis, where the canal angles at the West Basin, the original old H. H. Farnum stone mill, built in 1832, still stands. It was erected by Dr. Ball who died of pneumonia while supervising its construction. Its power came from diverted surplus canal water controlled through sluice gates directly into great water wheels, after which it coursed its way to the Delaware through Waste Weir Brook.

The waste weir that fed the mill, as well as the Clove Brook Waste Weir and the upper Waste Weir Brook, were the safety valves for Port Jervis in

Port Jervis about 1893 showing the sharp turn the canal made when leaving the Delaware Valley to head northeastward up the Neversink Valley. At the bend in the canal may be seen the old Farnum stone grist mill built in 1832. The canal to the left of the Main Street bridge was known as East Basin while that to the right side, West Basin. Note the large Erie roundhouse to the right. In picture D 28 a large section has been removed.

D 27 Collection of GERALD M. BEST - From WILL ALLERTON Glass Plate, GUS KRAUSS Collection

the control of high canal water. This corner of the Delaware and Hudson Canal was unique in that water flowed from two directions - down the Delaware section and from the north-east via the Neversink section.

In times of heavy rain, Port Jervis was in a precarious position.

Annoyance of the good people of Port with the canal company over occasional inundations was not the only thing that created strained relations. A foot bridge built by the canal company created considerable turmoil in 1857. It all started in August of 1851 when the foot bridge was erected over the canal at the north end of Pike Street (Orange Street on Port Jervis Map) to accommodate some of the forty dwellings on the west side of the canal. By 1857 the physical condition of the foot bridge was such that letters were arriving in prodigious number on the desk of the editor of the *Tri States Union* newspaper . . . "The fact is, Mr. Editor, the bridge is dangerous, and in no slight degree, to passengers. Many, particularly elderly persons, are driven to go way around to the Team Bridge for fear of life and limb. The accumulation of snow and ice is such as to make passage hazardous to the young and sprightly, let alone the aged and infirm, of which I am." **16**

D 28 Courtesy of J. A. O'BOYLE

View of the Erie Railroad yards showing the old suspension bridge built by the Erie to connect with Matamoras, Pennsylvania. This was also the great age of the Camelback locomotive as evidenced by the more than 15 shown at ease on the radial tracks in the foreground.

For all the complaining, the bridge still stood.

On the lighter side, Port Jervis became a center for excursion trips on the canal. In 1872 J. L. Burton outfitted a packet boat for excursion purposes and named it *Dyberry* after the headwater stream at Honesdale.

To publicize the new craft Mr. Burton invited 300 friends to accompany him on a trip to Butternut Grove, located between Port Jervis and Huguenot. On the way home on July 2, 1872, "A fife and drum corps made things seem lively and a most delightful time was spent by all on board. Captain Burton provided a bountiful supply of iced lemonade, and some of the guests brought large baskets, from which a pleasant supper was extemporized. The boat returned at 10:30 p. m. Two days later (July 4) saw the *Dyberry* doing a big business, carrying Independence Day visitors between Port Jervis and Sparrowbush . . .

" . . . On August 2, the Delaware Council No. 10, Order of American Mechanics, arranged for a grand picnic and excursion. The picnic was held in Butternut Grove and the excursion was made on the *Dyberry*. The Erie Band was engaged to discourse the music." **17**

And so it went all through the summer of 1872, one exhilarating excursion after another. D 29 and D 30

For the *Dyberry* and Mr. Burton it was not to last. The railroads quickly caught on to the idea and by the summer of 1873, by dint of their novelty, they quickly drew the customers away from the towpath and tiller.

The gradually declining coal tonnage on the canal, now coupled with the loss of excursion revenue, was but one more manifestation of the slowly deteriorating position of canal transportation.

The handwriting was on the wall.

D 29 Courtesy of ELLENVILLE PUBLIC LIBRARY

Even loaded coal boats were pressed into use for excursion use. The funnel shaped object in back of the man, left, is a bilge pump, known also as a pump-box hole.

D 30 Courtesy of J. VANCE HUNT

A Sunday School outing complete with brass band.

The Neversink Section

The Neversink Section ran from the Upper Waste Weir Brook at the northeast edge of Port Jervis to Neversink Lock #51 where the Neversink feeder entered the canal. As noted previously the flow of the canal on this section was against the loaded boats, a unique condition on the D. and H. Canal. This flow of water necessitated the building of four heavily framed timber stop locks: the Upper Waste Weir Stop or Guard Lock, Pine Woods Guard Lock, Hornbeck's Basin Guard Lock and the Huguenot Guard Lock.

The canal skirted the base of the mountains on the west side of the Neversink Valley, then under Issac Cuddeback's Bridge to swing around a sharp bend along an embankment above Culvert Brook. It was at this point that one of the largest breaks on the canal occurred, washing out over 8,000 cubic yards of earth through an opening 600 feet long.

The Port Jervis *Union Gazette* of April 2, 1869 reported, "When the canal broke, two empty boats which were on the way back were drawn through the opening. A force of about 300 workmen has been engaged in repairing it. In a day or two it will be in condition for light boats, although it will require a longer time to put it in complete order. The two boats that were washed out will be put back in the canal by the use of blocks and tackle. The canal will stand the cost of having them put back."

At Hornbeck's Basin the canal took a sharp swing through a right turn for its approach to Huguenot. Huguenot, with its boatyard, was at one time larger than Port Jervis, exceeding it in population, influence and importance. Its schools were considered the best in the Town of Deerpark. After the enlargement of the canal, new and larger dry docks and yards were erected there. The quality of the work turned out by Birdsall's Boat Yard was known up and down the canal. Today (1965) the boat yard area is a pasturage with

NEVERSINK SECTION

Port Jervis to Cuddebackville

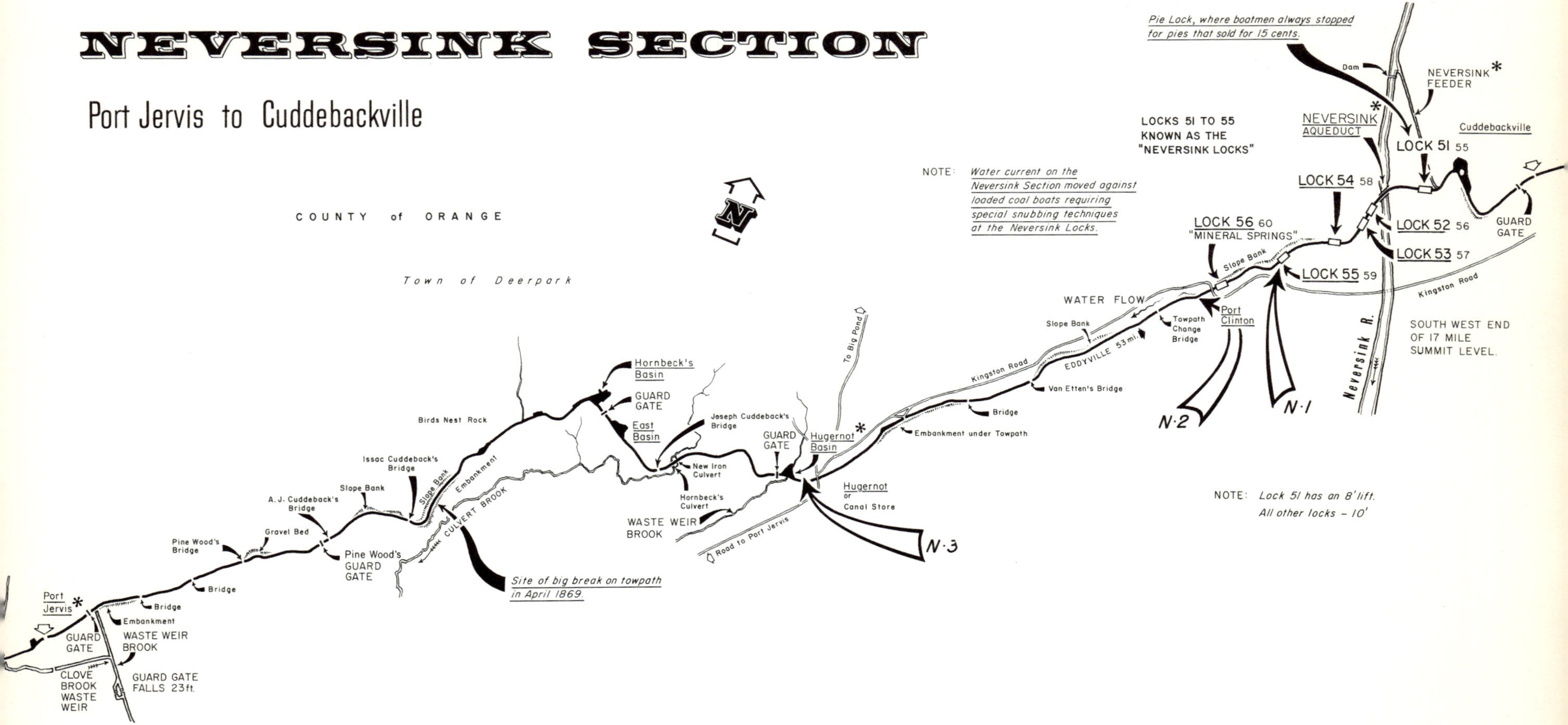

N 1 M. B. WAKEFIELD

Lock tender's house at Lock #55, the first of the five Neversink Locks. The ruins of the lock are directly in back of the house whose commanding location caused the road to be built around it years ago. The road has been maintained as such, down through the years.

traces of a foundation wall here and there. The larger of the two dry docks is a mere depression in the pasture while the stone work at the entrance gate to the smaller dry dock still stands. Debris fills much of the balance of this dock. N 3

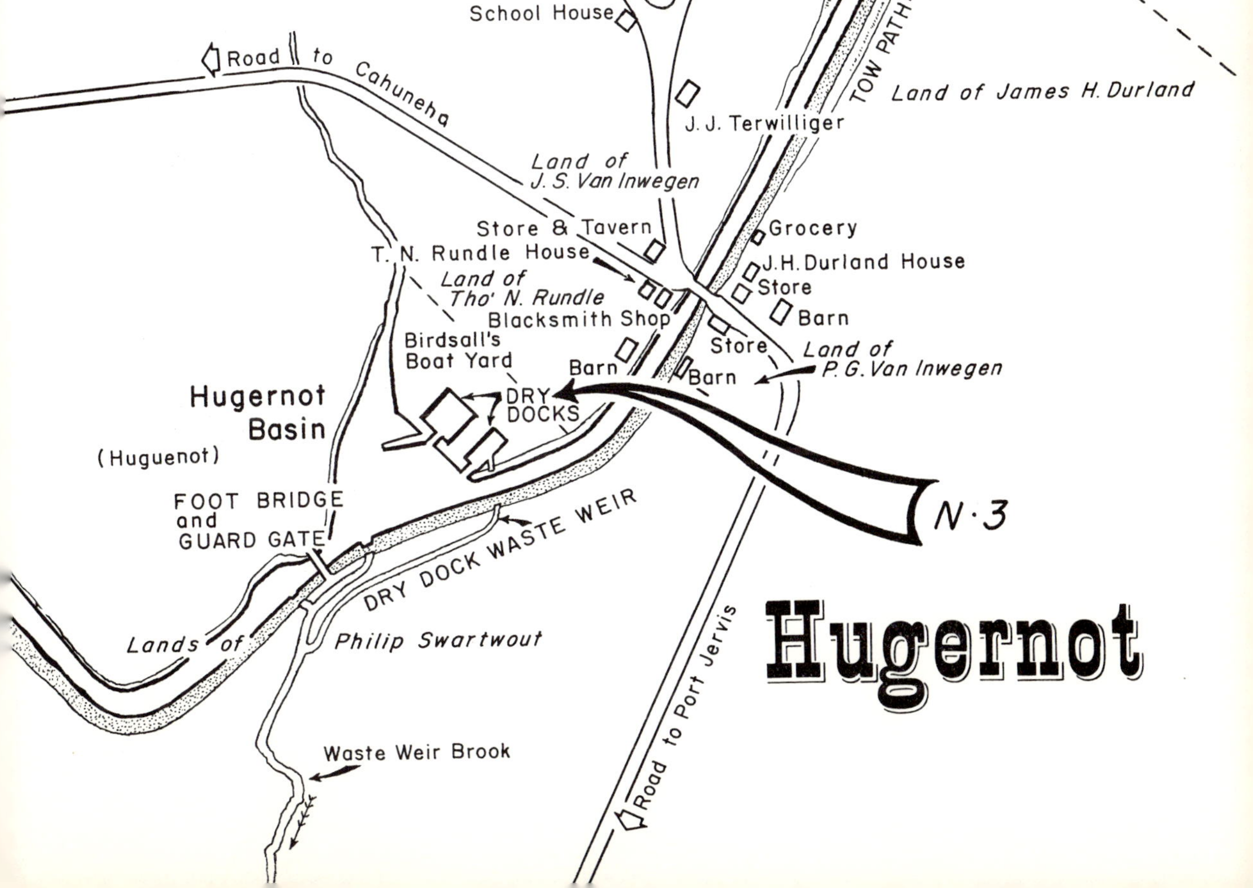

N 3

The remains of dry dock at the 'Hugernot' Basin. The right wall at the entrance to the dock is gradually being demolished since it has become a menace to the horses stabled in the pasturage. The depression of an even larger dry dock is out of the picture to the right.

In September of 1877, hard times on the canal prompted the company to issue orders to all boatmen to conclude boating operations if they so desired.

One of these boating families at the *Mineral Springs Lock* "was totally destitute of money, victuals and horse feed, and amid all these trials the wife gave birth to a child. Although their home was at or near Barryville, a wretch was found who refused to let their boat go by his so they could get home. They were thus compelled to await their turn on the canal, which would take them over one week as the boats were running. A man at Huguenot gave them all the food he had to spare, and thus relieved them temporarily. Other families were living on baked apples, and their teams on whatever they could pick." 1

The canal bridge at Mineral Springs Lock had at one time signs on both sides, one that read, "54 Miles To Honesdale" and the other, "54 Miles To Kingston". N 2

One of the typical lock tender's houses still standing (1965) may be seen at Lock #55, the first of the Neversink locks. N 1. Locks #51 to #55 were known as the Neversink Locks and were unique in that special ropes, known as the Neversink Locks Lines were needed to get loaded boats up and over the Neversink Aqueduct.

"This Neversink Locks Lines was a double line pully arrangement and was hitched onto the stern timberhead. After the towing line flew off when passing under the

N 2 M. B. WAKEFIELD

Looking southwestward down the canal from the Mineral Springs Lock, #56. A power company has taken over the towpath as a transmission line right-of-way.

N 4 Courtesy of ELLENVILLE PUBLIC LIBRARY

Abe Hoaz looking down into the lower gate end of Lock #51, also known as the Pie Lock. The ropes on the right were instrumental in the use of the Neversink Locks Lines and the control of the upper gate.

lock bridge, the driver would bring the towline forward and hook into the ring that wrapped over the rod that ran down to the lower gates. The horses would then tow the loaded boat against the slight current of the water into the lock. The boat snubbed and gates closed, it would then be raised up to the next level." **2**

N 5 Collection of MALCOLM BOOTH

The Pie Lock from the lower gate end with a repair boat passing through. The Neversink River feeder passed down along the base of the hill in the background.

The Neversink River is reached just after passing through the gaunt abutments of the abandoned Monticello branch of the New York, Ontario and Western Railway. On the left, in back of a trailer court, the ruins of Lock #53 may be seen at the base of the railroad fill. N 6. The now filled-in Lock #51, known up and down the canal as the *Pie Lock* is found just after the crossing of the Neversink River.

Moses VanInwegen and later his sons Job and Stoddard tended the Pie

Lock and store. It was at this store that Mary Casey manipulated the dough so expertly that canalmen waited in line to buy her bread. She also made little pies and treats for boys and girls who came by on the canal, hence the name Pie Lock. N 4 and N 5

At times altercations arose, disturbing the peaceful scene of canalmen waiting patiently in line for their bread and treats. "Wednesday morning James Schoonmaker and Anthony Sampson of Rondout, in charge of a boat passing through the Neversink Locks, got into a dispute with Mr. Calvin Griffin, the lock tender. The belligerent boatmen, both of them large heavy men, proceeded from words to blows. They both assaulted Mr. Griffin, who is a man considerably beyond the prime of life, and shamefully beat him.

"As soon as possible, Mr. Griffin hastened to Port Jervis, and warrants for their arrest were obtained and put into the hands of Chief Burton. Mr. Burton met the boat as it was passing the glass works in this place (Port Jervis), arrested the men, and escorted them before Justice Dutcher. They pleaded not guilty, but after stating their case to council, under his advise, they changed their plea to guilty. Each was fined $25." [3]

N 6 M. B. WAKEFIELD

Lock #53 of the Neversink Locks approaching the Neversink Aqueduct. This lock had a 10-foot lift as did all the Neversink Locks with the exception of Lock #51, which had a lift of 8 feet.

The massive abutments still stand at the Neversink River to mark the site of one of the four suspension bridges built by John A. Roebling. N 7. Close inspection of the southwest abutment will reveal the massive eye-loop anchors into which the stranded wire rope was wound and cinched up. N 10

In February of 1847, Roebling started preparation of a preliminary estimate of the Neversink Aqueduct. He estimated that a one-span aqueduct across the Neversink would cost $25,000, while a two-span structure could be built for $18,000, plus $500 for building an anchorage on the west face. [4]

N 7 M. B. WAKEFIELD

Looking toward the northern abutment of the Neversink Aqueduct. The four pylon-like stone towers held the cables after they emerged from their eye-loop and concrete footings.

N 8 Collection of MALCOLM BOOTH

View of the Neversink Aqueduct from above the tracks of the Port Jervis, Monticello & New York tracks at Rose's Point. The center arrow points to the old 1827 bridge abutment while the one on the right locates Lock #51 or the Pie Lock. The left arrow indicates the base of the hill along which ran the Neversink River feeder.

N 9 Collection of MALCOLM BOOTH

Towpath-level view of the Neversink Aqueduct from the south anchorage, showing a whirling post along the left edge of the picture.

N 11
Courtesy ELLENVILLE PUBLIC LIBRARY

The feeder dam across the Neversink River from which the water is delivered down a long feeder to the canal adjacent to Lock #51. The open type building housed the intake gates.

N 10 M. B. WAKEFIELD

A cluster of six massive eye-loops emerge from the downstream abutment of the Neversink Aqueduct.

The single span proposal was adopted. The span measured "from center to center, 170 feet, with pyramids, two feet higher than those on the Delaware Aqueduct; bottom of the saddle above the canal bottom, 14 feet, seven inches; point of intersection above the canal bottom, 14 feet, seven inches; point of intersection above the canal bottom, 14 feet, 10½ inches; camber, 11 inches; cables above beams at the center, three and a half inches; cables below bottom at the center, one and a half inches; lowest point of cables above canal bottom, nine and a half inches; deflection of cables, 14 feet, one inch." [5]

Work on the new suspension aqueduct was started in June of 1849 just below the old canal crossing. N 8. The new bridge was suspended by two 9½ inch cables, consisting of 3,024 wires each. The aqueduct was completed in the winter of 1850 - '51 in time for the 1851 boating season. N 8, N 9 and N 12

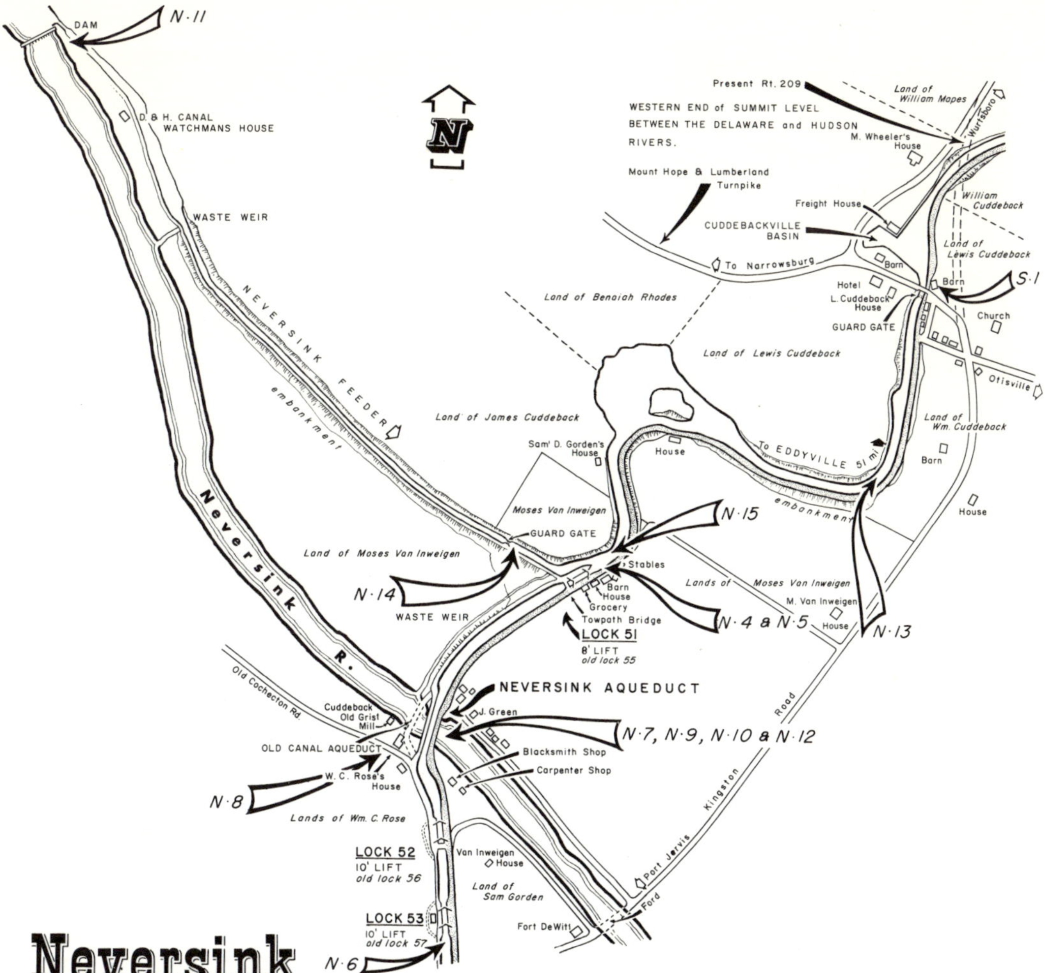

Neversink Aqueduct and Cuddebackville

N 12

N 14

The ruins of the Guard Gate on the Neversink Feeder. This barrier retains water in the old feeder back to the Neversink.

N 15

The small bridge in the center marks the point where the Neversink River feeder enters the canal. The area to the left is the site of Lock #51 at the southwestern end of the 17-mile lock-free Summit Level.

On the other side of the Neversink Aqueduct at Lock #51, the Neversink feeder permitted much needed water to flow into the canal for the Summit Level. N 15

A large dam was built across the Neversink River upstream to carry the water into a long feeder ditch. N 8 and N 11. This ditch still contains water for a good part of the distance to the canal, being held back by the remains of the old feeder guard gate. N 14

N 13 M. B. WAKEFIELD

At Cuddebackville the canal winds around the Cuddebackville Cemetery, the Towpath and water held in place by a high embankment.

Additional reservoirs were developed in the hills west of the Summit Level to supplement the waters from the Neversink. Even with the extensive reservoir systems of Wolf's Pond, McKee's Pond and Lord's Pond, droughts did plague operations all too frequently on the 17-mile level.

Beyond Lock #51 the canalboats moved out along a high embankment to begin the long haul down the lock-free Summit Level to Phillipsport. N 13

The Summit Level

From Lock #51 the canal right-of-way contains water through the basin at Cuddebackville and beyond to Comfort's Basin, creating a summer home and recreational area. For about two miles this section suggests the days when the canal was in operation. S 1 and S 2

Because of the absence of locks it was popular in the summertime for excursions and outings. These excursion events took place on converted canalboats, special steam launches and even revenue-bearing canalboats. D 29 and D 30

The Port Jervis *Evening Gazette* reported an unusual excursion from Cuddebackville to Wurtsboro. "At 8 o'clock a. m. delegations began to arrive from Huguenot, Oakland, and the vicinity of Cuddebackville. At 9 o'clock the word was given 'all aboard.' About 100 invited guests took seats aboard. Word was given to go ahead, when the slack line was taken up by two beautiful teams, and we darted through the waters of the canal at a rapid rate. On arriving at Westbrookville we stopped 10 minutes, where we were joined by several of the prominent families of that place. The order to go ahead was given by our Mate, James Mitchell, Esq., and again we were moving rapidly through the beautiful valley of the Mamakating. We stopped a few minutes at Brownsville, where we were joined by other friends. Arrived at Wurtsboro at half-past 12 o'clock. A procession was formed under the direction of Marshal M. L. Cole, and the party marched to the hotel of Samuel Gumaer, where we were well assured that we were welcome. We sat down to a sumptuous dinner prepared by Sam and Lady, to which I must say our party did ample justice. The short time in Wurtsboro was agreeably spent by calling on persons and receiving calls. At half-past 4 o'clock the party returned to their boat, and left amid cheers and good-bys for Cuddebackville." [1]

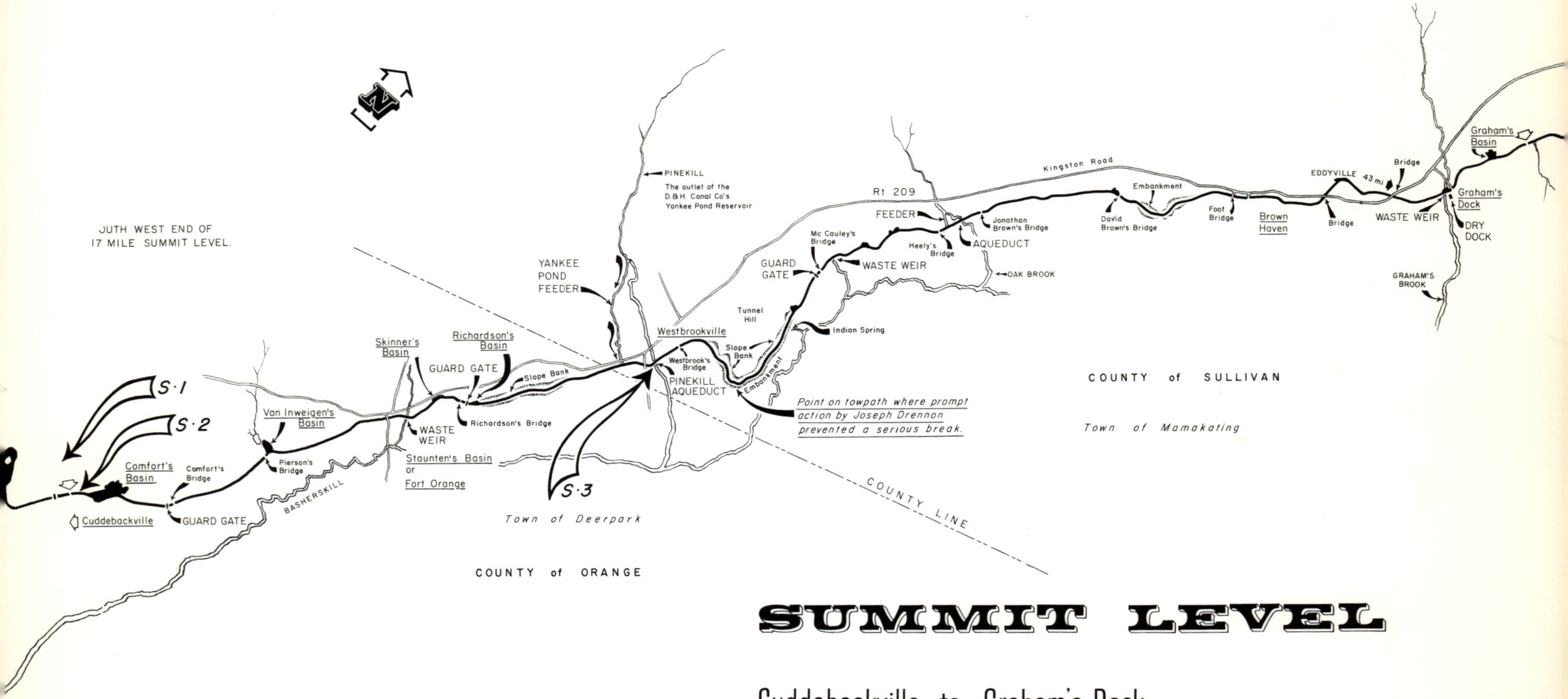

SUMMIT LEVEL

Cuddebackville to Graham's Dock

Graham's Dock to Phillipsport

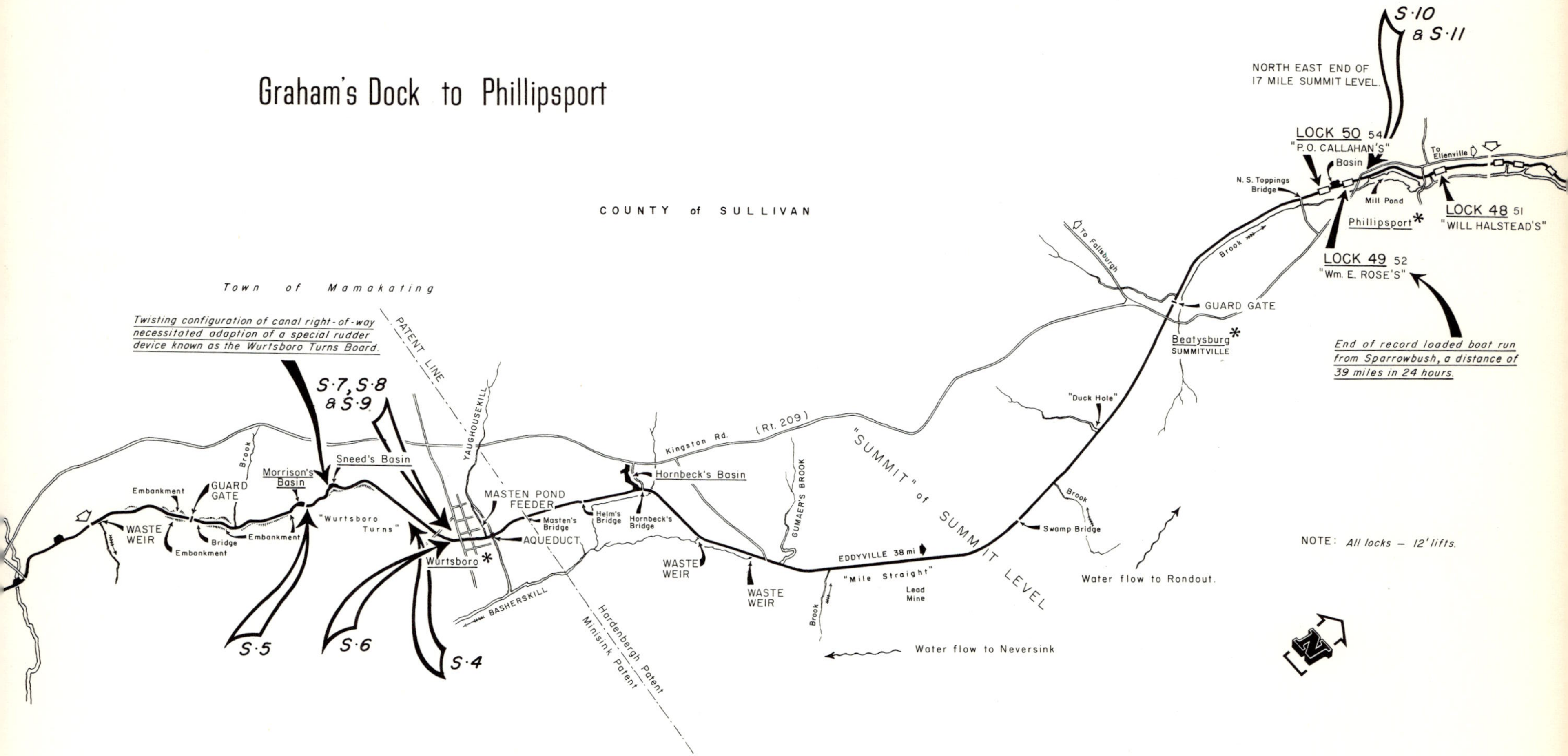

S 1

Looking across the once bustling Cuddebackville Basin. A freight house once stood where the chicken coop is located on a former boat slip. Directly in back of the photographer next to the towpath are the foundation remains of a stable.

At Westbrookville the Pinekill feeder from Yankee Lake entered the canal just south of the Pinekill Aqueduct. Very little remains of the Pinekill Aqueduct which was located just below the big 'S' turn around Tunnel Hill. S 3

Two or three miles west of Westbrookville were some very valuable bluestone and building-block quarries. Large quantities of customized stonework were shipped out on the D. and H. Canal from the Westbrook area to distant points. [2]

S 2 M. B. WAKEFIELD

At Cuddebackville the canal right-of-way contains water for about two miles. This scene is from a point just south of Comfort's Basin.

As may be noted on the Summit Level map, feeder streams were few and far between. This lack of adequate feeders and paralleling water courses, coupled with subsoils of an unstable gravelly composition, resulted in frequent periods of low water. "Boating on the canal has again been discontinued on account of the want of water. Since the late rain storm about two boats have been brought down, and half of their cargoes disposed of. There are now several hundred boats detained at the Summit Level." [3]

S 3 M. B. WAKEFIELD

Remains of the Pinekill Aqueduct at Westbrookville, feeder stream from the Yankee Lake reservoir. These ruins are located just a few hundred feet east of Route 209.

There did occur periods of abundant water supply however. In the summer of 1884 a Sullivan County correspondent reported that the Delaware and Hudson Canal Company had not yet felt the need to draw upon the reservoirs which were well above the high water mark.

Frequently, when the boats would be held up on the Summit because of low water, the idle boatmen would start burning fences, robbing gardens and orchards, causing the canal-side farmers to kick up quite a fuss. Most boatmen respected the lands of the farmers but there was also the very tough and irresponsible type who respected nothing, including their own boats and equipment.

"The greatest evil of all was the reckless usage of the boats. In most instances, young, vulgar, and depraved men were employed to run and manage the boats . . . whose chief delight arose from the destruction they could cause to the property of the canal company. Hence they would wreck and tear the boat cabins, deposit in them all manner of filth, write upon the walls the most vulgar sentences, draw the most rude obscene forms and pictures, tear the siding and roof from the cabin and the lining from the boat until nearly every line boat on the canal became a most unseemly object to behold. They would in many instances run the boat in their charge against rocks thereby causing it to leak and many times to sink." [4]

Canalside dwellers had to maintain a constant vigil to protect their property and womenfolk. As John Johnston further related, "It was annoying, vexatious and in a large degree burdensome when it was not only necessary to guard every implement or article of personal property with utmost vigilance but they could not shield the sensibilities of their families . . . from the tumultuous sounds and vulgar voices which boatmen, most of all, could issue. A woman or girl could not pass a boat or scarcely become visible around her house and escape the mortification and shock to her modesty at all times held in store for her." [5]

The canal-delinquents also had their fun throwing rocks at locktenders or passers-by, some using improvised slings for greater range and accuracy. At other times armed with long poles they would harrass unsuspecting occupants

Looking southwestward toward the 'Wurtsboro Turns' from the Otisville Road bridge at Wurtsboro. The 'turns' start where the canal bends to the left in the distance. Note the piles of cordwood racked up along the towpath on the left.

S 4

Collection of M. B. WAKEFIELD

of privies perched along the canal's berme edge. This cruel and hilarious sport was made possible by the silence of motorless, slow navigation and the general disrepair of privies. [6]

During the 1880's the dropping freight rate on coal was reducing the canaller to starvation wages. Finally the point was reached where wives and children were used to help navigate the boats. On occasion they were run without any helper. In spite of these existing conditions the men did not strike. An earlier attempt at striking, back in 1862, resulted in retaliation on the part of the management who "sank two or three boats and threatened to sink others if they continued, the boatmen concluded it was best to resume navigation to avoid loss of their boats." [7]

The practice of the company in building boats and selling them to the boatmen also became subject to question, particularly when the cost to the company was pared down to eleven hundred dollars. "Despite all pretentions at vigilance, the construction of boats was of the cheapest material and the most hurried workmanship. Such boats could be contracted to an individual for $1900, eight hundred over and above its cost to the company. Added to that, fifty dollars was charged the purchaser for the high privilege of getting a new boat, and, as though the extortion was not yet sufficient, mileage was charged and regulated by the distance the boat lay from Honesdale, the place of loading when the boatman took possession." [8]

In order to realize a reasonably rewarding season on the canal, the boat operator had to get in 13 to 15 trips. Frequently he would be denied this because of a 'stop the boats' order that would go out simply to control the market price or out-put of coal. As Johnston relates, "A telegraphic order would run the length of the canal to the local managers, 'Stop the boats', and thus the boatmen were subjected to an arbitrary delay, the duration or course of which they were ignorant of. Thus they might be detained one week or time sufficient for an entire trip; with no allowance to lessen their burden.

S 5 Courtesy of PENNSYLVANIA HISTORICAL AND MUSEUM COMM.

A typical whirling post turn on the Delaware and Hudson Canal. Two whirling posts may be seen on the towpath in the distance.

When the secret purpose was accomplished, the order would be wired from Honesdale, 'Start the boats'. By such scheming process the trips of the boatman would number not more than 8 or 10 trips in the season." [9]

The boating season could also be seriously interrupted on the Summit Level by breaks in the towpath. In April of 1872 a break two miles north of Cuddebackville carried out 70 feet of the towpath and washed out the bottom of the canal to a depth of 18 feet. A force of 140 men were promptly put to work and the job completed in just four and a half days.

Watchmen were hired by the canal company to be on the alert for conditions in the towpath that could lead to a serious break. The Port Jervis *Evening Gazette* of April 11, 1878, reported the prompt action of " . . . Joseph Drennon, a watchman on the canal at Westbrookville, who while patrolling his beat Wednesday, discovered a leak through the towpath at Tunnel Hill, at a place where the embankment is very high. With the aid of several neighbors he succeeded in putting up the stop lock by which the water was cut off. Had it not been for the prompt action of Mr. Drennon a serious break would have resulted, as the bank at that place is composed of coarse white sand through which the water was rapidly making its way."

The canal wound northwestward from Tunnel Hill, over Oak Brook where a small feeder entered the waterway. Graham's Dry Dock and Basin were next, after which Morrison's Basin and the famous *Wurtsboro Turns* were encountered.

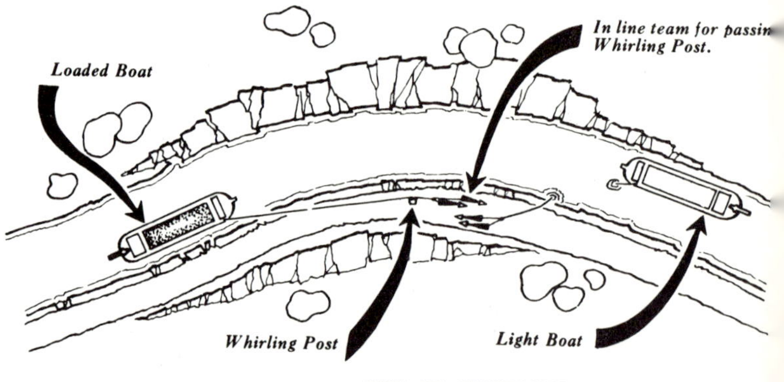

USE OF WHIRLING POST

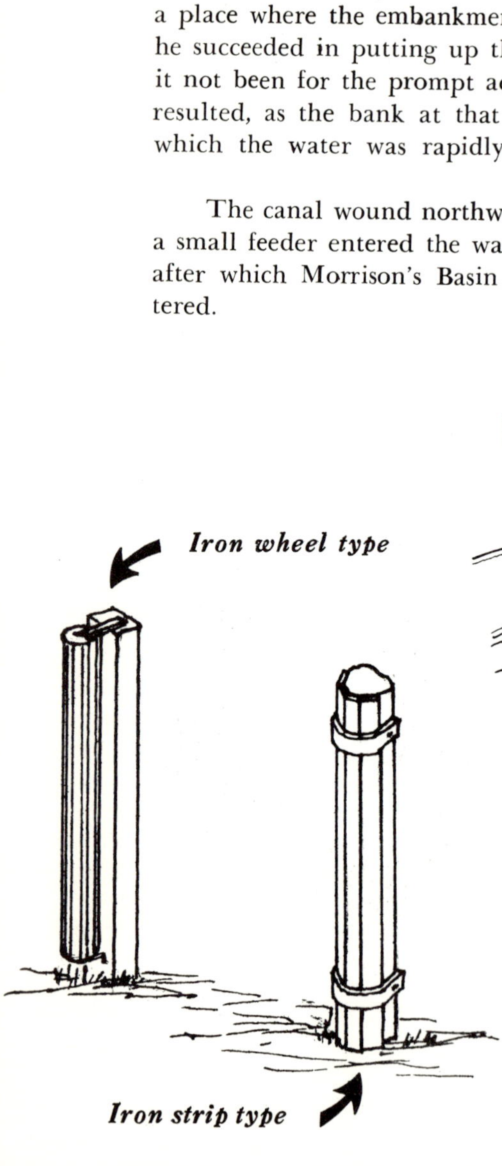

It was at locations such as the Wurtsboro turns that *whirling posts* had to be used. S 4, R 30 and N 9. Whirling posts stood about five feet in height and had an iron wheel that ran down one side. Another type used consisted of iron strips secured down the canal side of a post, and attached with two strips of horizontal band iron, top and bottom.

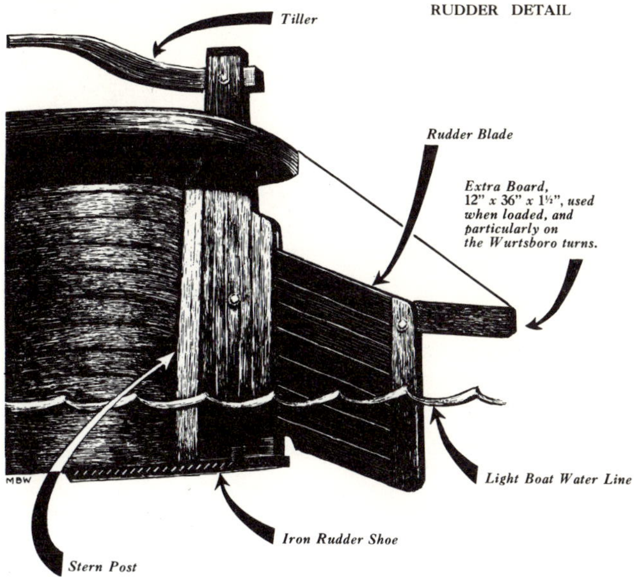

RUDDER DETAIL

According to DuBois Weber, "When a light boat would meet a loaded boat at the Wurtsboro Turns (S 4) the light boat would pull over to the inside or berme side and allow its ropes to settle to the bottom, the horses moving to the outside of the whirling post. The loaded boat team would then pass single file between the whirling post and the water, towing the loaded boat safely over the light boat lines. The whirling post kept the loaded boat line from drawing against the light boat team."

An additional aid to negotiating the Wurtsboro Turns was a small board that stuck out from the top of the rudder blade called the Wurtsboro Turns Board. (See diagram right) This board was also of help when the boat was carrying an unusually heavy load of coal.

James A. O'Boyle explains it as being hinged on a bolt through the top of the rudder blade which could be flipped back by a small rope leading to the tiller. This was particularly necessary when locking through. The rudder blade itself fitted into a slot in the bottom half of the rudder post. This pivoted on a bolt through the rudder post. When the boat was light this blade dropped down below the water for steering. When the boat was loaded the blade, being made of wood, floated up to clear the bottom.

Under normal passing conditions the light boat would take the towpath side, the loaded boat the berme side.

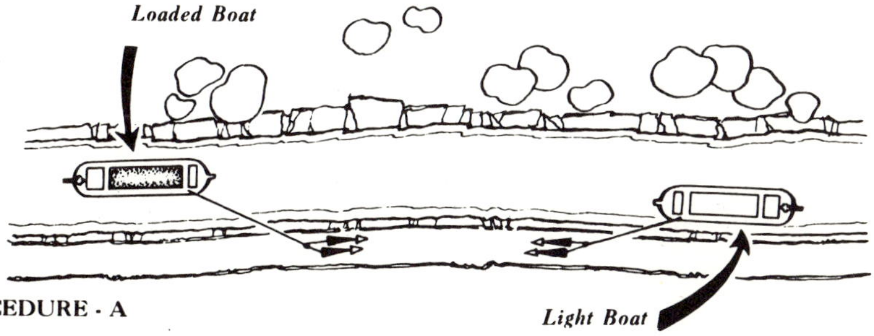

PASSING PROCEDURE - A

The loaded boat team would draw its boat until the lines came up alongside the light boat team.

PASSING PROCEDURE · B

The loaded boat team would then stop to permit the light boat team to pass over its line. Both teams would continue to tow until the loaded boat was within a few feet of the light boat; then they would stop the loaded boat, allowing its towline to settle to the bottom of the canal.

PASSING PROCEDURE · C

The light boat would then pass safely over the loaded boat's line. If the line on the loaded boat was new they would have to stop sooner to allow the line to settle to the bottom, a new line taking longer to settle than one which was old and waterlogged.

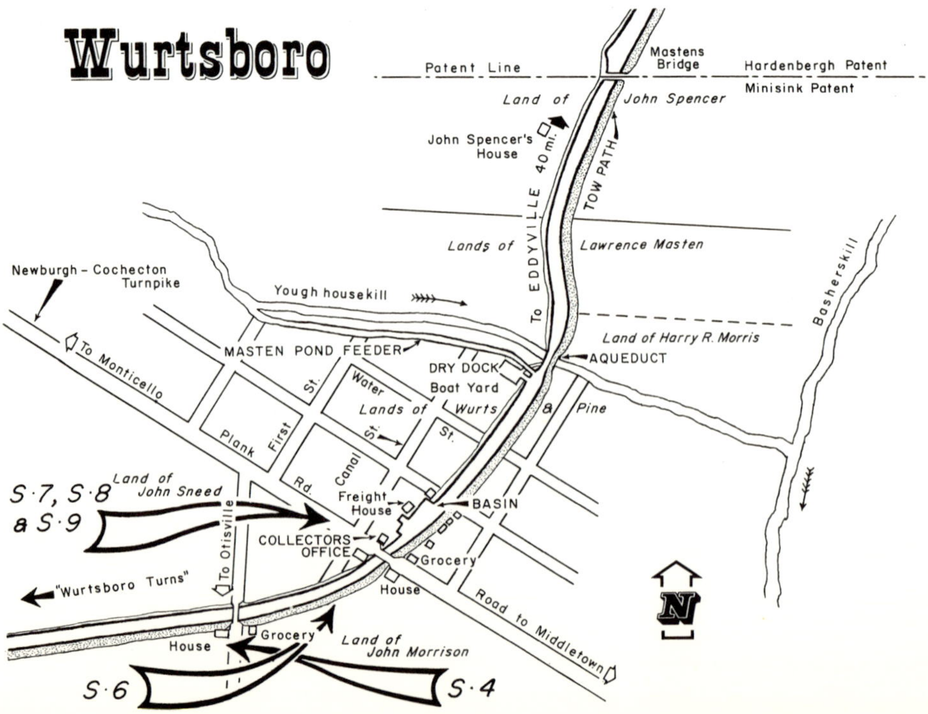

S 7 Courtesy of MRS. THOMAS MOORE

This classic by an itinerant photographer shows the Seaside Supply Store in Wurtsboro enhanced with an assortment of the community's civic leaders. Note the bespangled musician with the cornet, probably on his way to a concert. This structure, somewhat altered after the canal closed, still stands.

S 6 Collection of M. B. WAKEFIELD

A light boat just south of the Newburgh-Cochecton Turnpike bridge in Wurtsboro. The D. and H. Collector's Office, managed by Charles G. Bennett as telegraph operator (1872), was located on the other side of the bridge. The building over the stern of the boat is the Seaside Supply Store shown in S 7. Note the little girl riding the mule.

At Wurtsboro the Masten Lake Feeder entered the canal after being diverted from the Yaughousekill at the northern edge of the village. Before the D. and H. Canal was built, Wurtsboro was called Rome, but upon its completion it was named in honor of Maurice Wurts. [10] S 6, S 7 and S 8

It was here that the canal was crossed by the important Newburgh-Cochecton Turnpike. This pike was instrumental in the opening up of Sullivan County and the development of its tanning industry. Raw hides from as far away as Argentina were transshipped at Wurtsboro.

S 8 Courtesy of MRS. THOMAS MOORE

The famous Harding House stood on the north side of the Newburgh-Cochecton Turnpike in Wurtsboro. This landmark across the turnpike from the Seaside Supply Store was a favorite stopping place for canalmen. Note the blurred image of a passing rowboat in the foreground.

In 1872, Wurtsboro contained three churches, a district school, three hotels, eight stores, an upper-leather tannery, a grist mill, a saw mill, two wagon shops, three blacksmith shops and about 650 inhabitants.

Jacob Stanton, superintendent of the D. and H. Canal feeder reservoirs in 1872, made his home in Wurtsboro, as well as Jacob Whitman, keeper of the Yankee Pond feeder reservoir. The D. and H. Canal night watchman in the Wurtsboro area was Harrison H. Stanton, and the foreman, William Harding. Boatbuilders Graham and Holmes had their works located next to the Masten Pond feeder. In 1872, eighteen boatmen made Wurtsboro their home port. [11] At the Harding House, located directly along the towpath, a thirsty boatman

S 9 M. B. WAKEFIELD

The old Harding House still stands today (1965) abandoned and in disrepair. The pole is part of a transmission line that follows the towpath.

found a handy port of call. Note in picture S 8 the convenient location of the bar room door to the towpath.

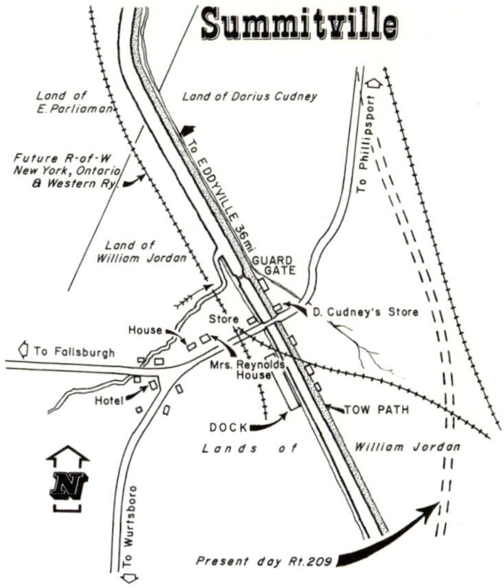

Just north of Wurtsboro was located Masten's Bridge which marked the *Patent Line* between the historic Hardenburgh Patent and the Minisink Patent.

A few miles beyond was the Gumaer's Brook feeder where the canal curved into the *Mile Straight,* the longest straight section on the canal. This straight section was also the precise summit of the Summit Level; from this point the waters drained north to the Rondout and south to the Neversink.

At Summitville (Beatysburg on official 1854 D. and H. Canal maps) the canal passed under the O. and W. tracks before swinging into the last mile and a half of the Summit Level. The Ontario and Western Railway bridge over the canal was, according to the Sullivan County Directory of 1872 - '73, "a fine work of art." Summitville in the same directory listed a schoolhouse, a sole-leather tannery, a store, 120 inhabitants, but no churches. 12

Lock #50, *P. O. Callahan's,* marked the end of the Summit and the beginning of the Ellenville Section. Lock #49, known to John A. McGinnis as *Rose's Lock,* was the end of a record run made by Captain M. McCardle in May of 1887. S 10 and S 11

S 10 Courtesy of ELLENVILLE PUBLIC LIBRARY

Bill Rose's Lock, officially Lock #49, as it was at the time of the record run by Capt. M. McCardle. The small lock tender's house on the towpath still stands and is used as a barn.

The Kingston *Freeman* reported that McCardle "left Sparrowbush (on the Delaware above Port Jervis) at 12 o'clock noon . . . one day and towed to Bill Rose's Lock at noon the next day, a distance of 39 miles. He claims that it is the quickest time ever made between two places with a loaded boat." [13]

This could be done only on this particular section since there were no closed locks for the night to impede the progress of the boat.

The ease of boating on the lock-free Summit Level, however, was offset by the concentration of nine locks in the Phillipsport area, the beginning of the Ellenville Section.

S 11 M. B. WAKEFIELD

The remains of Bill Rose's Lock as it appears in 1965.

The Ellenville Section

At Phillipsport the lock builders had a field day. It was here that the boats commenced their gradual descent to the Hudson River. Phillipsport was a bustling boat building canal town in 1873 with one church, a school, one hotel, two stores, three groceries, three boatyards, one wagon shop, two blacksmith shops, three shoe shops, one millinery and two dressmaking establishments, a saw mill and about 400 inhabitants. E 1

The leading boatbuilders at Phillipsport were Wesley B. Snyder who had his works at Lock #45, Charles Medler, and Russell & Baker who operated a dry dock and boatbuilding yard. At Lock #42 James A. Caldwell operated a boatbuilding yard while his brother Stephen built and repaired boats at Lock #40.

In 1873 Rodolphus S. Smith was the superintendent foreman in Phillipsport for the canal company while William G. VanBuren listed himself as a canal carpenter. Oddly enough, the 1873 directory fails to list a single boatman as living in Phillipsport. The growth of this village was based strictly on the repair and building of boats.

The company officially listed Phillipsport as having ten locks. This is true if those north to Spring Glen, which is about a mile north of the boatbuilding town, are also counted. In any event, starting at Lock #50, *P. O. Callahan's,* they progressed northward, like steps, through the village in a straight line to Lock #46. Just beyond Lock #46 the canal followed the stream which acted both as a feeder and a waste weir in the village area, into a right-hand bend to Lock #45.

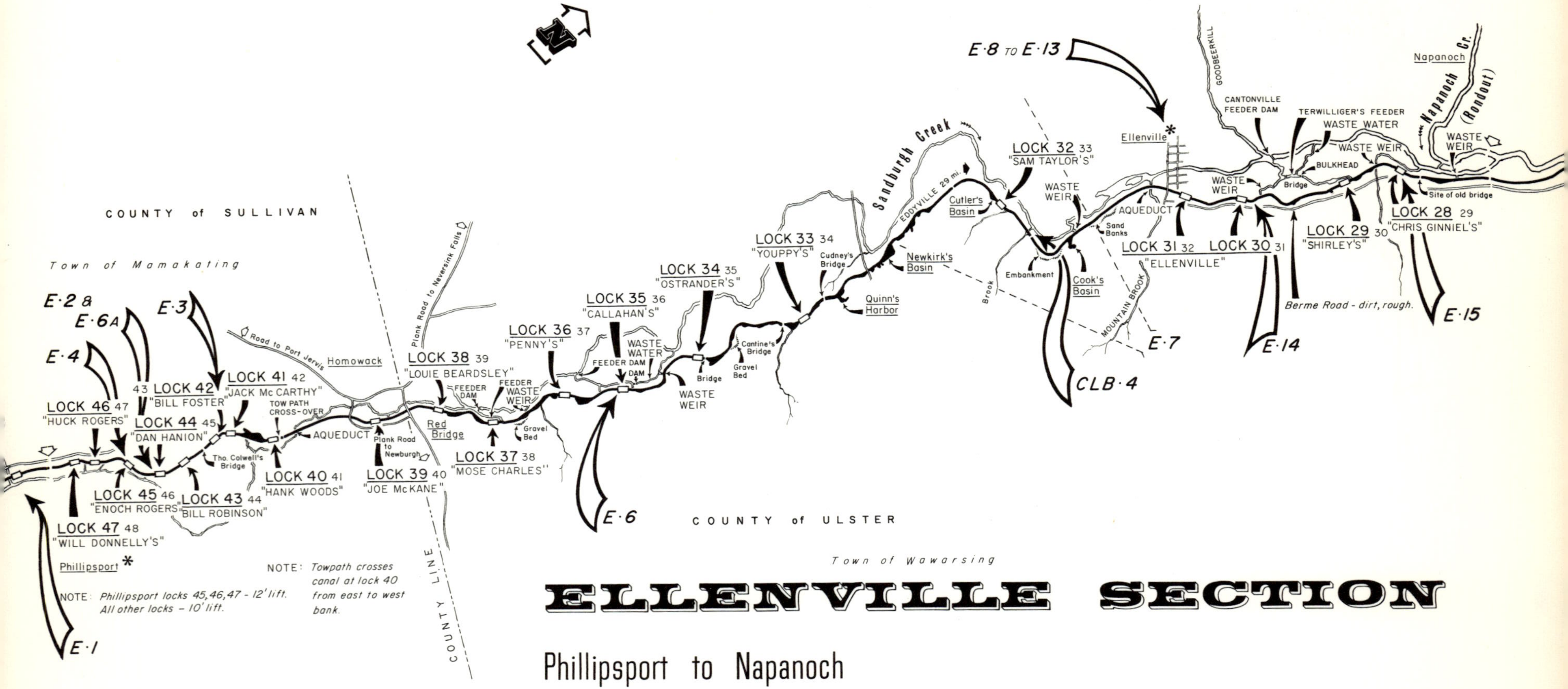

ELLENVILLE SECTION

Phillipsport to Napanoch

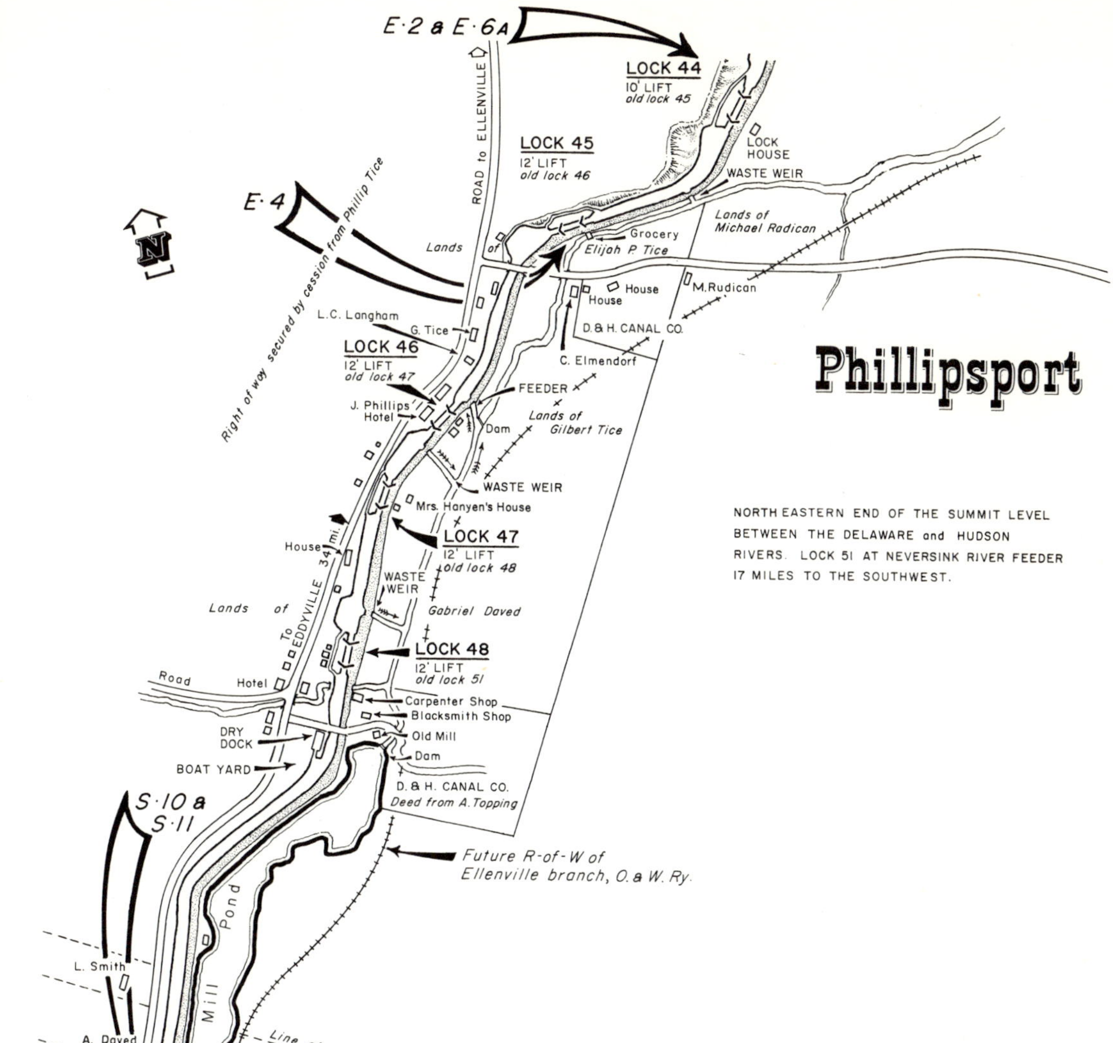

E 1 Collection of M. B. WAKEFIELD

The canal just north of Lock #49 (Rose's Lock) at Phillipsport showing the Mill Pond on the right. A waste weir may be seen on the towpath by the pole in the foreground. Present day Route 209 runs directly on the canal and towpath at this point.

E 2 M. B. WAKEFIELD

Scale model of Lock #44, prepared by the author for clarification of the proposed restoration of a series of locks in the Phillipsport - Spring Glen area. *(See E 6A)*

In August of 1961 a prospectus was prepared for the development of a *living** Canal Museum and Park area, to be created out of the canal remains, starting at the basin south of Lock #45, and thence northward through Locks #45, 44, 43, and 42. E 2, E 3, and E 4. A proposed concept was evolved that would show each of the two types of locks used, with a third lock left in its 'primitive' state. E 3

The late New York State Historian, Albert B. Corey, and Mr. Abraham George, Jr. of the New York State Conservation Department journeyed to the site to meet with the author and take a walking survey of the proposed restoration. E 5. Lack of sufficient water and the high restorative costs negated the project.

Just north of Lock #40 was located a towpath cross-over from the east to the west bank. This was done to bring the path alongside the water courses it would be following for the balance of the journey to High Falls. (Detail, Page 173)

Homowack was the last locking point for loaded canalboats in Sullivan County and was also the home port for the noted canal freighters W. and I. Cudney. Nearby at Lock #39 was located the large canal store operated by George Newkirk. The *Tri-States Union* of October 25, 1878, reported this store "totally destroyed by fire together with its contents (early Wednesday morning)

*** "living" i.e., one in which the full-scale boats and locks would be operational.**

E 3

Both Photos by PAUL GERRY

The passing basin at Lock #42. It was proposed in 1961 to keep this lock in its primitive state. To the right was once located the boat building yards of James A. Caldwell.

E 4

The author at Phillipsport Lock #45 pointing out a badly eroded towpath. This lock was one of four included in the 1961 proposed restoration. Note the protruding outside segment of the lock gate recess chamber at left, a construction detail normally covered by the earthen fill of the towpath. At one time a canal grocery store was located over the cascading waters.

E 6 M. B. WAKEFIELD

Callahan's Lock #35 and its delapidated tender's house. A small canal store business was also conducted here. These ruins lie at the end of a dead-end road and in 1872 were the responsibility of Lock Contractor Virgil Spearks of Homowack.

E 5 Photo by PAUL GERRY

On August 15, 1961 three of the principals involved in the proposed canal restoration made a walking survey of the site. They are left to right: the author, Manville B. Wakefield, Sullivan County Historian; the late Albert B. Corey, New York State Historian; and Mr. Abraham George, Jr. representing the New York State Conservation Department.

E 6A

Detail of scale model of Lock #44 showing the lower gate house. Overflow waste weir passes in back of shanty.

which consisted of a large stock of dry goods and groceries, and a small amount of mail matter, the Post Office of Homowack being located in the building."

After making the very wide reverse curve through what is now the grounds of the Nevele Hotel, the canal swung gently through a graduated curve into Ellenville. E 7. Locks #31 and 30 were known as the Ellenville Locks although Lock #30 was located quite a distance from the populated part of the village. A very busy drydock, located adjacent to Lock #31, was operated by Mr. Hoar in the closing years of the canal. The basin which extended from the Center Street bridge to Lock #31 was the heart of canal activity in the village. E 8 and E 9

Facing the basin and adjacent to Lock #31 was the extensive canal-side store of S. A. VanWagner. E 10. Today (1965) it is a secondhand furniture store and a printing shop. E 11. Also still standing is the former Merchant Tanners Line warehouse. E 12. The Delaware and Hudson Canal Ticket Office which once faced the drydock has since been relocated on Brodhead Street. E 8

Extensive slip facilities were built just north of Lock #31 for the Ellenville Glass Works and the Brodhead Company and warehouse. Ellenville, unlike Phillipsport, had its share of boatmen; the 1872 - '73 Directory of Ulster County listed ten in residence. [1]

Just north of Ellenville, Lock #28 was tended by Albert Genegal who commuted daily from that village.

E 7 Collection of JIM SHAUGHNESSY

View of the canal from the Shawangunk Mountains, just south of Ellenville, showing Sam Taylor's Lock #32 in the distance. The Nevele Hotel complex today covers the entire area to the left of the canal. Note the newly laid O. and W. tracks in the foreground.

E 8 Courtesy of ELLENVILLE PUBLIC LIBRARY

View of Ellenville Lock #31 from the Center Street bridge showing the collector's office just to the right of the lock. The small house with the columns to right of center was the D. and H. ticket office, later turned around when the canal closed.

E 9 M. B. WAKEFIELD

Today (1965) the Ellenville boatbasin is a short street with three canal-era buildings still standing. The large building at right was the warehouse of the Merchant Tanners Line which operated six boats in local freight to New York City.

E 10 Collection of JIM SHAUGHNESSY

The Ellenville Basin looking southwest showing Lock #31, left. A light boat may be seen about to pass under the Center St. bridge. Note the boat loading cargo at the Merchant Tanners Line warehouse, left, and the omnibus parked just beyond the S. A. VanWagner Store, right.

E 11 M. B. WAKEFIELD

Today Lock #31 has been filled in, the main east-west highway Route 52 taking its place. All of the buildings shown here once faced the towpath.

Ellenville

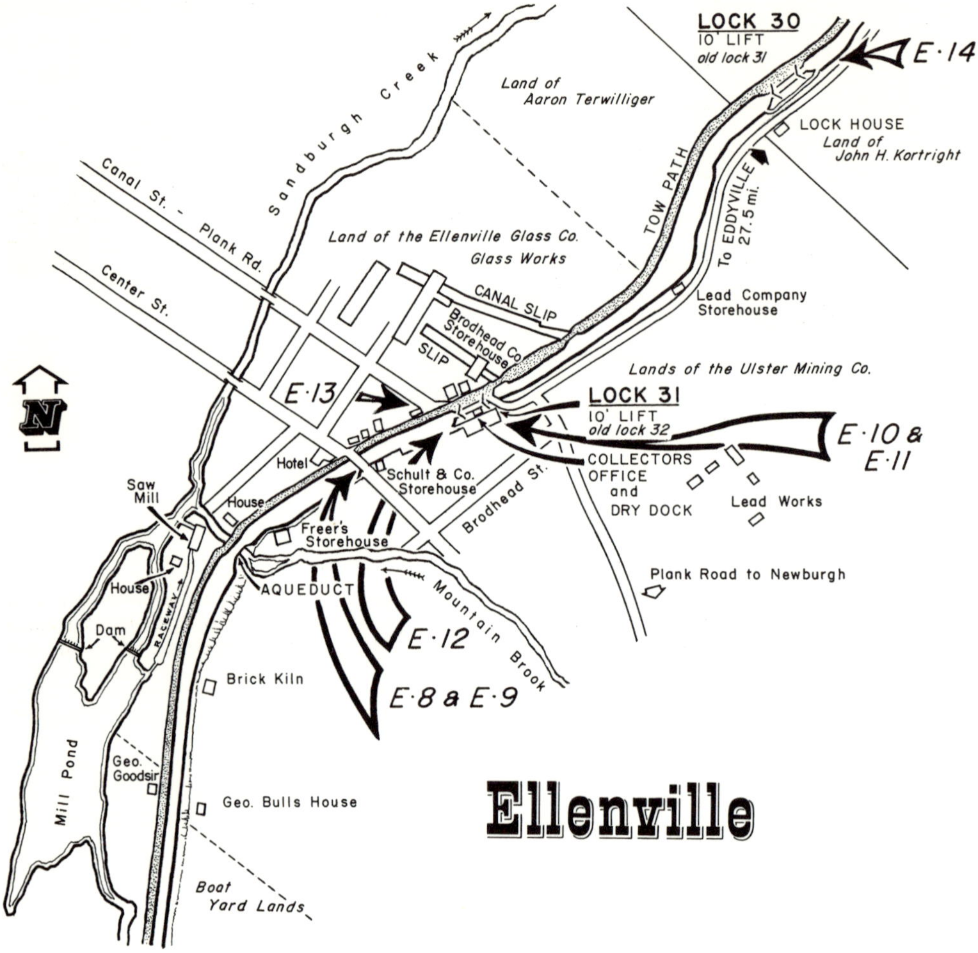

A loaded boat passing a light boat on the approach to Ellenville Lock #31. The light boat is moored at the warehouse dock of the Merchant Tanners Line with the Hoar Drydock in the right background.

E 12 Courtesy of ELLENVILLE PUBLIC LIBRARY

E 14 M. B. WAKEFIELD

The weed-grown Ellenville Lock #30 just north of the village with the Berme Road to the left.

The Berme Road starts at Ellenville and runs northeastward to Kerhonkson. This dirt road is wedged in between the old canal remains and the precipitous flank of the Shawangunk Mountains. Much of the canal remains may be seen from this road in the spring, fall and winter. The summer season is not recommended for canal ruin investigation because of the dense foliage of trees and underbrush.

Just north of Lock # 28 the Sandburgh Creek joins the Napanoch Creek to form the Rondout Creek and marks the beginning of the Rondout Section of the Delaware and Hudson Canal. E 15

E 13 Courtesy of ELLENVILLE PUBLIC LIBRARY

A light boat in front of Lock #31 showing the old Kimble house next to the Canal Office. Hoar's drydock was located directly in front of these buildings.

E 15

Lock #28, also known as Ginniel's, (Genegal's) at the northeast end of the Ellenville Section. The dilapidated condition of the lock gate-house and the wagon tracks on the towpath would indicate the canal is no longer in operation.

The Rondout Section

"About one hundred men are employed in repairing damages to the canal between Ellenville and Napanoch, under the supervision of Mr. George S. Kimble."

So read the lead story of the Port Jervis *Evening Gazette* of January 2, 1879.

The storm which struck on December 9, 1878, became a torrential downpour by the 10th, and inflicted the heaviest damage ever experienced in the Rondout Valley. Opposite Napanoch, where the Napanoch Creek enters the Sandburgh, the canal completely disappeared along with bridges and buildings. Eight full breaks occurred between Ellenville and High Falls, tearing away thousands of feet of towpath and flooding the adjacent countryside. The damage was most severe at Eddyville and will be touched on later.

The canal curved around the Eastern Correctional Institution at Napanoch for which much of the stone and building materials had been brought in on the canal. R 1 & R 4

At Port Ben, later called Wawarsing, was located the canal store, operated by Mr. A. S. Clyne, two drydocks, and two extensive boat yards. R 3

Port Hyxon [1] was named after William Hickson who operated an

R 1 Line drawing by M. B. WAKEFIELD

***A light boat approaching the Eastern Reformatory for Men* (1898) *shown in the background. Note the small bridge on the berme and towpath side that carried the canal over a dry wash. The remains of these abutments are noted by arrows in picture R* 4.**

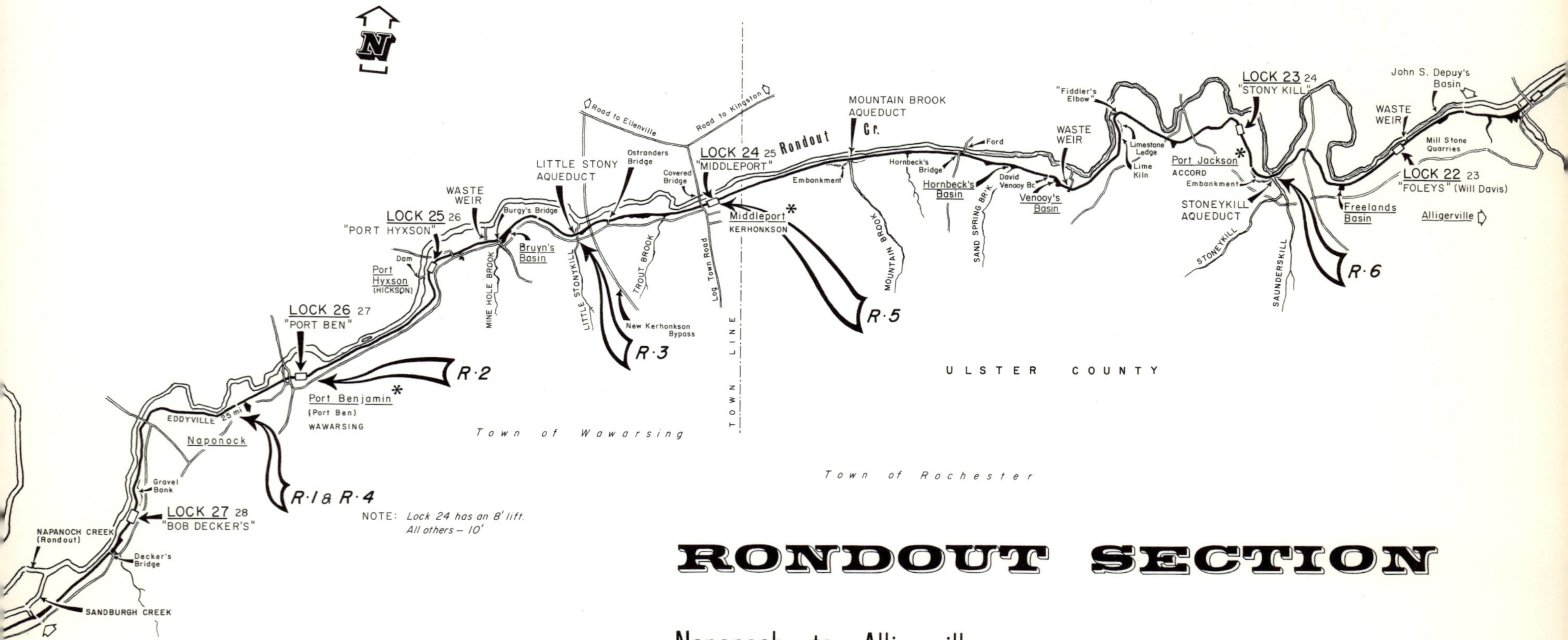

RONDOUT SECTION

Napanoch to Alligerville

Alligerville to Eddyville

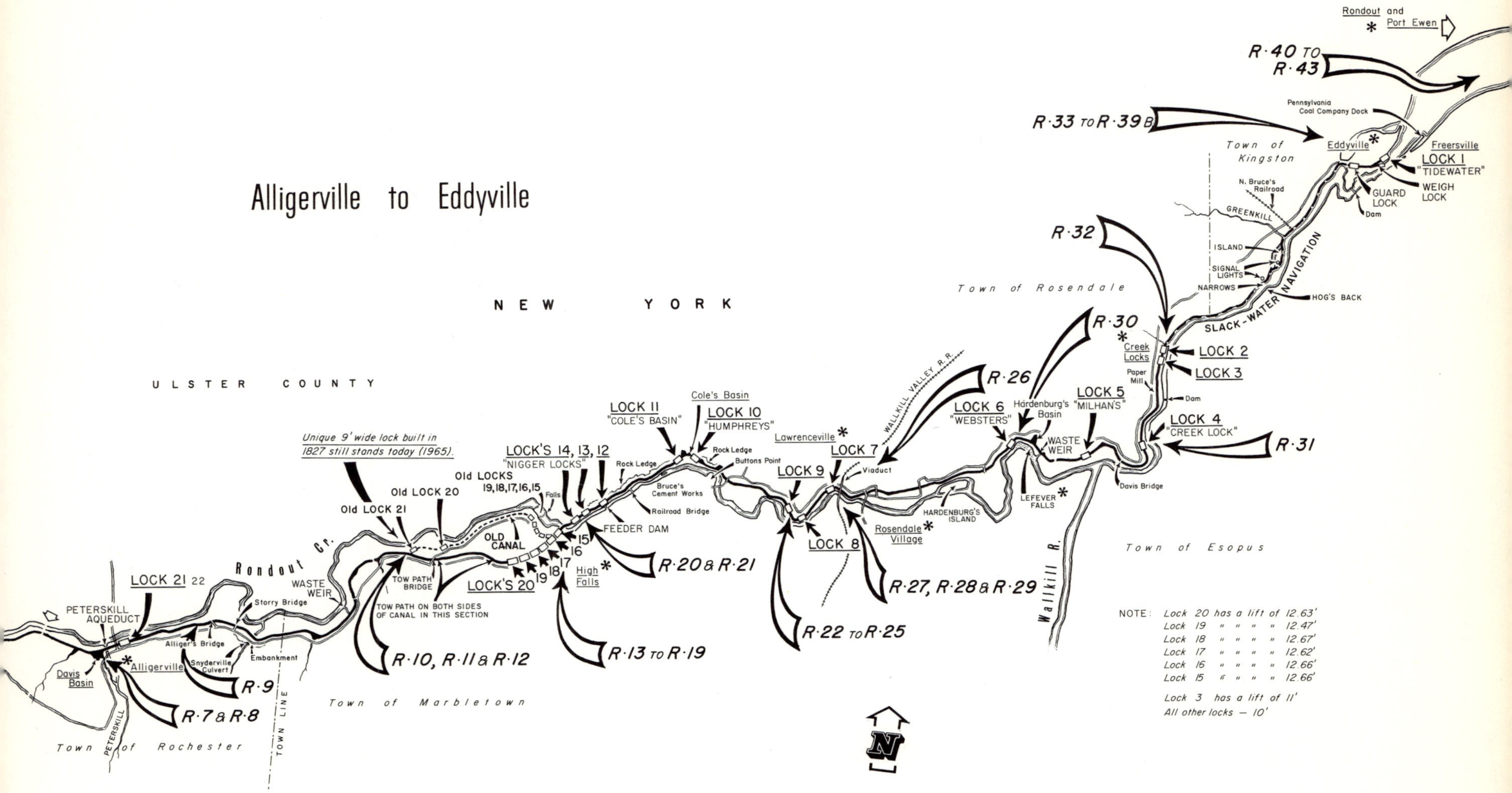

extensive farm at the spot in early pioneer days. Today just a few stones stand to mark the location of Lock #25. Down the Berme Road, just before the new Kerhonkson bypass, the ruins of the Little Stony Aqueduct are discernable on the left. R 3

R 3 M. B. WAKEFIELD

Site of the Little Stony Aqueduct south of Kerhonkson. The low wall on the right held the flume-like wooden trunk that contained the canal water and boats. The far abutment once held the steel beams of the small bridge used by the O. & W. Railway.

The canal passed through the heart of Kerhonkson, then known as Middleport, with Lock #24 situated at the spot that was later to become the location for the New York, Ontario and Western Ry. station. R 5. In 1872 four boatmen operated out of Kerhonkson: William W. Addis, Augustus Churchill, W. H. Decker, and Richard Schoonmaker. Strangely enough for a community of its size, there were no dry docks or boat building activities. [2]

The Berme Road ends at Kerhonkson with an option of two routes from there to Accord: Route 55 to the turn for the Granit Hotel and thence down to Accord, or Route 209 north. [3]

R 2 Courtesy of ELLENVILLE PUBLIC LIBRARY

The A. S. Clyne canal store and stables at Port Ben (Wawarsing). The chains lying along the edge of Lock #26 in the foreground were used by the lock tender to control the action of the upper gate (left) from his position over the lower gates (out of picture at right). On most locks this action was controlled by chains lying on the bottom of the lock.

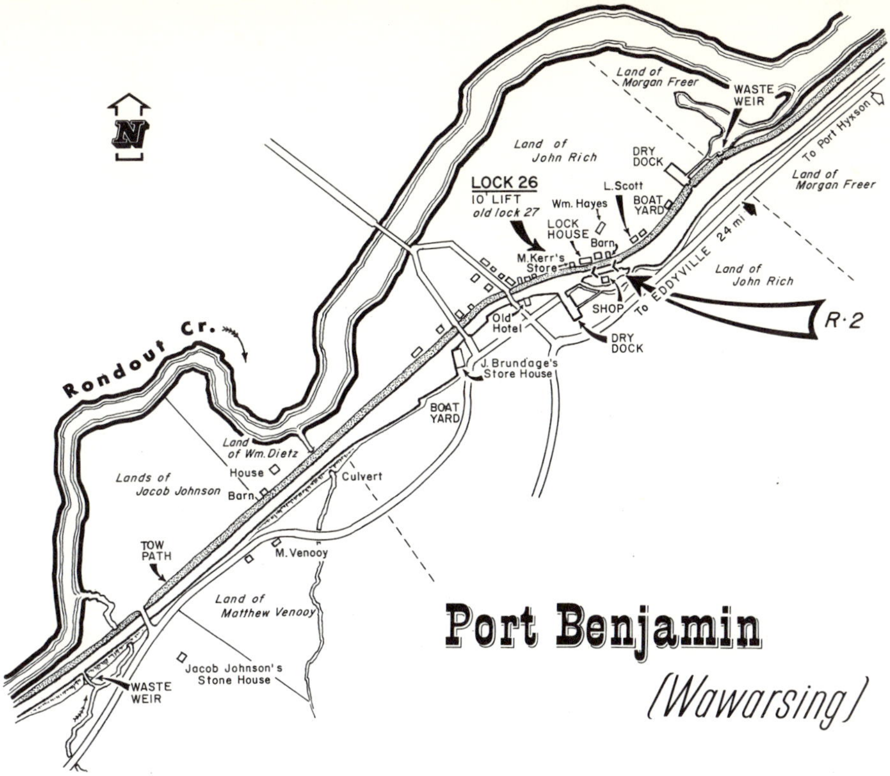

A road that roughly follows the canal picks up again at Accord, originally known as Port Jackson, where two rather extensive basins were once located, Hardenburg's and Schoonmaker's. The canal north of Accord swung out on an unusually high embankment for the aqueduct crossing of the Stonykill. R 6

James S. McEntee, in his story of the early canal days written in 1874, recalls the well-authenticated story of the vision of an elderly gentleman many years prior to the canal's existence. The old man was crossing from the settlement on the east side of the Stonykill to the settlement on the west side via a tree which formed a bridge, when he saw coming toward him from the opposite side, a boat whose deck was crowded with well dressed ladies and gentlemen. "They had music and were dancing, and a festal air pervaded his apparent vision. He became alarmed at the unwonted spectacle, and returned to tell the incident to

R 4 M. B. WAKEFIELD

The abandoned canal right-of-way from the same view point as that in picture R-1. Scattered railroad ties may be seen on the right on the old Kingston branch of the O. and W. Ry., once the towpath. Arrows indicate small bridge abutments.

Middleport

(Kerhonkson)

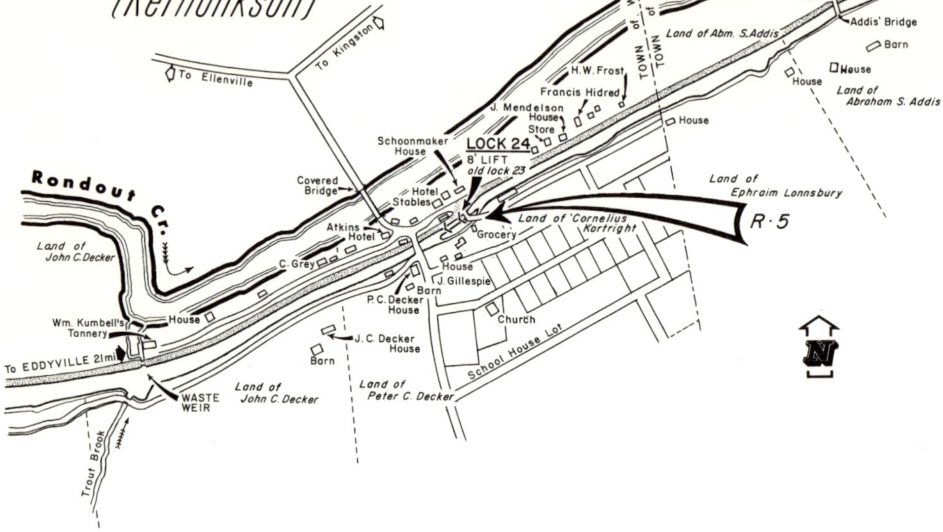

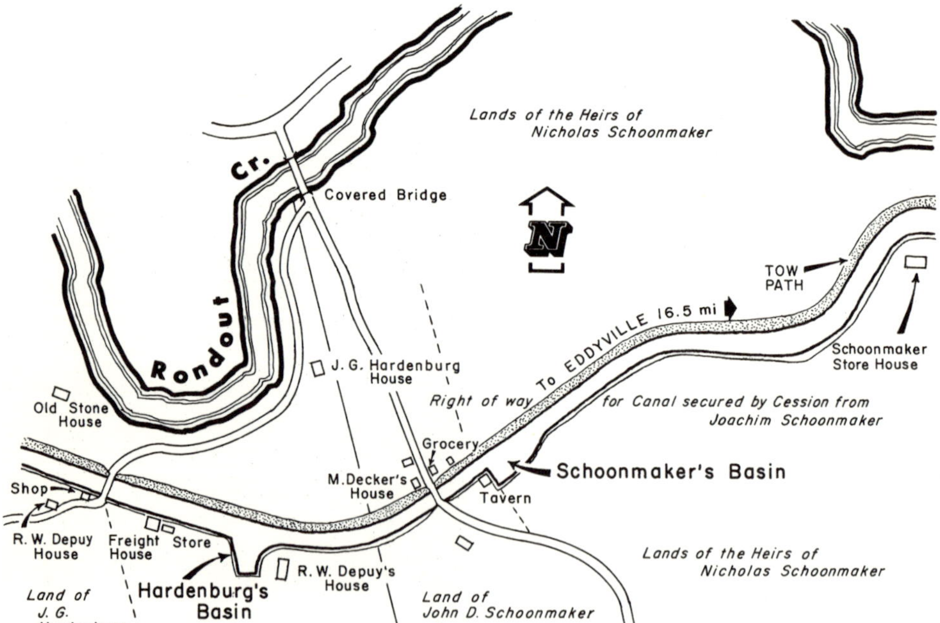

Port Jackson *(Accord)*

R 5 Line drawing by M. B. WAKEFIELD

Lock #24 at Middleport (Kerhonkson) showing a loaded boat snubbed in, waiting for the upper gate to be raised. The second building down on the right was a canal grocery which, along with the lock, became the site of the O. & W. Railway depot.

his family and friends, giving them a minute description of the boat passengers and the direction in which it came. The canal as surveyed and located was almost identically in the same course of his phantom boat across the flats." [4]

The road swings a considerable distance away from the canal until the Peterskill is reached at which point it turns to follow the stream to Davis Basin at Alligerville. R 7

Davis Basin was located just south of the large double-trunk Peterskill Aqueduct. R 8. The road crosses the still operational small feeder that led into Davis Basin. A sharp right turn across the berme side of the aqueduct leads into the village of Alligerville. [5]

R 6 M. B. WAKEFIELD

The northern abutment ruins of the Stonykill Aqueduct just north of Port Jackson (Accord). The approaches to the aqueduct were carried on a high earthen embankment.

A bluestone loading dock, dry dock, boat yards and Lock #21 marked the route of the canal through Alligerville, north of which it wound through gently rolling terrain. Today the residents of the area show

R 7 M. B. WAKEFIELD

Davis Basin just south of the Peterskill Aqueduct at Alligerville. The canal curves gracefully down through the woods on its way to Accord. It was named in honor of Peter Davis, father of Washington H. Davis. "Wash" Davis used to walk, day after day for many years, along the towpath to his general store a mile away in a building occupied today by Mr. and Mrs. Christiana.

pride in its presence by maintaining it and keeping it clear of refuse. R 9

In Chapter Two it was noted that when the canal was deepened and widened, a segment near High Falls was built on a new right-of-way. This fact was noted on the official 1854 maps of the canal company. On the Rondout Sectional map the old route is shown as a dotted line with two locks, #20 and #21, indicated on the southerly end of the bypassed section. At the time of the author's survey of this portion of the canal, old Lock #21 was found to be in remarkably good condition. R 10, R 11 and R 12

This lock is unique in that it is the only surviving lock of the canal as originally built in 1827 to a width of nine and a half feet. Old Lock #20 has virtually disappeared as well as most of the pre-1850 canal towpath and berme that is still vaguely discernible winding along the river. Progress, in the guise of a recent highway re-alignment in High Falls, wiped out old locks #15 through 19.

In 1852, 6 the new series of locks, numbered 15 to 20, were built at High Falls to better compliment the new Roebling Suspension Aqueduct alignment. DuBois Weber remembered talking to an old-timer who recalled how the new locks were built. "The huge

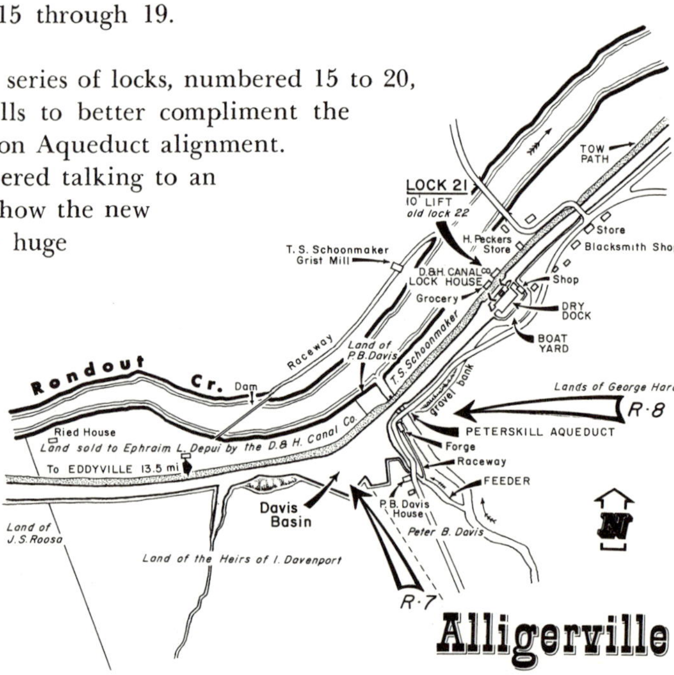

R 8 M. B. WAKEFIELD

The Peterskill Aqueduct abutments just south of Alligerville. A highway bridge is built on the berme side while a foot bridge has been built across the towpath abutments. Floods have taken out the berme side of the center pier.

R 9 M. B. WAKEFIELD

The canal at Alligerville with the towpath on the left. A short distance north the only remaining original 1827 lock may still be seen where the new 1850 alignment veers to the right, away from the river and the old right-of-way.

stones in the lock were not put up from the inside but were built from the outside. They put a big four foot block of stone on the bottom and then brought up the berme side and towpath side even with the stone. Then a stoneboat would slide up, pulled by two or three yokes of oxen with a big stone on board. This would be pulled up on the earth filled-in in back of the first four foot stone, the stone slid off and set in place. And so it went; they laid a row at a time and filled in back with dirt until the lock wall was finished with towpath and berme done at the same time." [7]

On February 13, 1863, the Port Jervis *Tri States Union* reported that a heavy slide of earth occurred at High Falls just above the first lock in the new cut, filling the canal for

Old Lock #21 between Alligerville and High Falls as seen from the lower gate end. The extremely narrow 9' 6" width of the locks, as originally built in 1827, marks this lock ruin as unique among D. and H. Canal artifacts.

R 11

R 12

A close-up of the recess in the wall of Old Lock #21. This cavity held the lock gates so as to permit the passage of a boat when the gates were opened.

R 10 M. B. WAKEFIELD

The old lock is located on a small knoll a bit higher than the Alligerville-High Falls Road and cannot be easily detected by the passerby. Unfortunately the upper end of the lock is being used as a dumping area.

over three hundred feet. "It is supposed that it will require the services of 150 men to remove the obstructions at that point, and to get the canal ready for business by the first of April." It was the proverbial ill wind blowing some good, in that it gave employment to the local laboring man during the season when it was most needed.

The series of locks at High Falls were precision cut and fit, with Lock #16, located at the old DePuy Hostelry, still in remarkably fine condition. R 13

R 13 M. B. WAKEFIELD

The well preserved Lock #16 at High Falls is built of cut and fitted stone. The large stone to the left is the keystone removed from the original 1826 stone arch bridge.

R 14 Courtesy ELLENVILLE PUBLIC LIBRARY

View of early High Falls from a stereopticon of the lock-dominated village. To the left-center edge is the telegraph office (1) and the Collector's Lock, Lock #15. Just beyond is the basin below the wagon bridge at the DePuy Hostelry (2). Locks #16 (3), #17 (4) and #18 (5) may be marked by the small shack over the lower gates. Some of the old stone work on the original 1827 canal right-of-way at Old Lock #15 shows in front of the building in center of the picture (6).

High Falls and Aqueduct

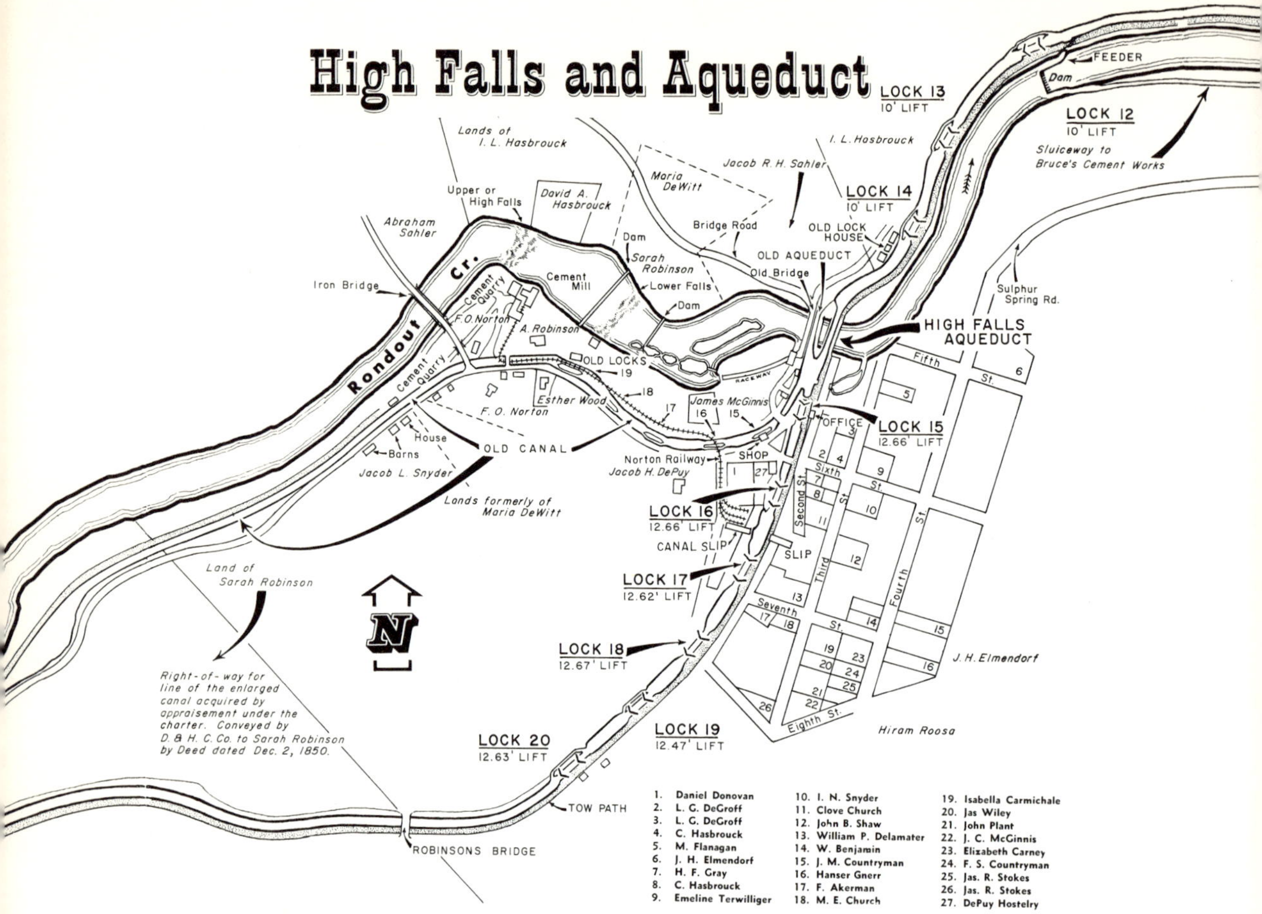

The lack of wood lining in these locks, each with a lift of 12 feet plus, created hazardous conditions as cold weather approached at the end of a boating season. DuBois Weber told of how "the water would squirt out from between the cut stones over the boats as you were lowered in the locks. As you got further down to the last lock the boat would be coated with ice with piles of ashes spread all over everything."

In 1872 the little community of High Falls contained two churches, two hotels, five stores, two meat markets, a grist mill, extensive cement mills and about 400 inhabitants. It was also known as a center for the shipping of bluestone flagging. Below the wagon crossing, (R 15) the canal widened into a basin at the Collector's Lock and Telegraph Station. R 14 and R 16

As originally built, the canal crossed the Rondout at High Falls on a double-arched stone bridge which, because of its rigid construction, could not be widened during the rebuilding program of 1847 - '52. R 17

John A. Roebling was completing the final stages of the Delaware and Lackawaxen Aqueducts when he started the High Falls Suspension Aqueduct in April of 1849, with the contract for the extensive relocation going to the firm of Story, King and Fuller. [8] In 1921 metallurgists dug some of the original wires out of the abandoned suspension bridge abutments for laboratory tests. The tests showed a higher tensile strength than originally reported by Roebling. Although exposed for 72 years to all weathering conditions, the wrought iron wires showed no material corrosion.

The suspension bridge disappeared after the closing of the canal, but the stone arch bridge withstood the ravages of floods and weathering, only to be demolished by an electric company. R 18 and R 19. It was, in jurisprudence terms, classified as an "attractive nuisance". It would seem that a cyclone fence erected around each end would have preserved a unique relic of a bygone era for generations of historians and students to come as well as, becoming a meaningful tourists attraction for High Falls.

The keystone of one of the two demolished arches is now on display at Lock #16. R 13. In the keystone was found a shallow copper box, five inches square, containing three early nineteenth century coins and three inscribed silver plates. [9]

As noted previously, High Falls played a key role in the pioneer years of the cement industry. During the summer of 1825 James S. McEntee, assistant

R 15 Line drawing by M. B. WAKEFIELD

The wagon bridge crossing at High Falls as seen from the Collector's Lock. The venerable old DePuy Hostelry, in the center, still stands today (1965) *next to the filled-in approach to Lock #16.*

R 16 Courtesy of ELLENVILLE PUBLIC LIBRARY

High Falls Lock #15, known as the Collector's Lock, with the telegraph office at right. The High Falls Aqueduct may be seen between the lock and the telegraph office, with the three Nigger Locks just beyond. The old canal, by-passed in 1848, ran to the left on the other side of the towpath directly in front of the buildings. This picture was taken from the wagon bridge over the canal near the DePuy Hostelry.

engineer for the D. and H., noticed rock outcroppings that resembled cement rock on the farm of Jacob L. Snyder. When this was brought to the attention of Jervis and Wright, specimens were burned in a blacksmith's forge at nearby Bruceville, reduced to powder by pounding, and tested by actual use. It was found to be of excellent quality, making the importation of Chittenango Cement from Madison County unnecessary. Mr. Littlejohn contracted to furnish all the cement needed for the canal in the spring of 1826. The first quarrying, burning and grinding operations took place in a kiln operated near the Sulphur Spring below High Falls by Simon DePuy.

Tight wagon boxes were used to haul it in bulk to the place where it was to be used and then shoveled into sheds. Mr. McEntee soon determined that the DePuy mills were insufficient to keep up with the demand which resulted

R 17 Line Drawing by M. B. WAKEFIELD

The two High Falls crossings show in this line illustration, the pioneer twin arch bridge built in 1826, and the Roebling suspension span commenced in 1849.

R 18

The original stone arch bridge at High Falls, from the upstream side, shortly before its demise.

in the building of more extensive facilities. With the completion of the canal contract, Mr. Littlejohn closed the High Falls works which lay dormant for many years until developments in the Rosendale area reactivated the industry.

R 19 M. B. WAKEFIELD

The demolition of one arch reveals the northern abutment of the Roebling Suspension Bridge. The bridge was so well built that the wrecking ball could not break the arch. Dynamite was used as a last resort.

Because of the crossing of the Rondout at High Falls, a towpath cross-over was located a mile or so south of Lock #20, so as to bring the towpath between High Falls and Eddyville on the outside or river side of the canal. (See sketch of typical cross-over bridge.)

TOWPATH CROSSOVER

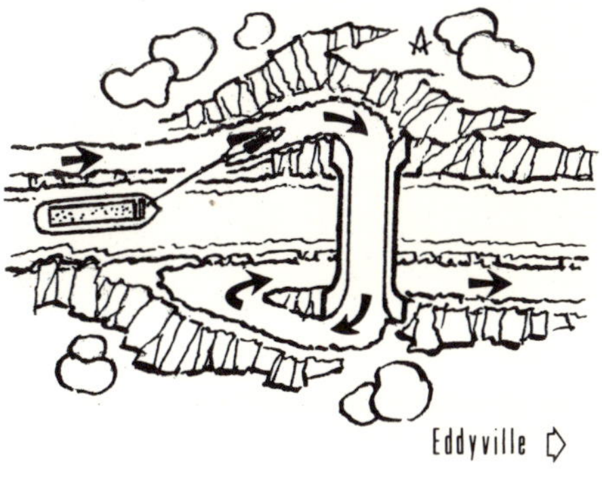

Just across the Rondout were located the three *Nigger Locks,* #14, #13 and #12. Locks #13 and #14 are still intact even to an occasional lock beam. R 20 and R 21

In 1826, cement rock was discovered in the Rosendale area. A man named Lucas Ellendorf, owning land in what was later to become Lawrenceville, made an arrangement with Watson E. Lawrence for the manufacture of cement. The venture was so successful that the village was named in his honor.

"Mr. Lawrence built his kiln small and used wood for burning. At the bottom of the kiln an arch was formed to contain the wood. The kiln was filled with stone after which a fire was kindled and kept burning for six days and nights when the stone was supposed to be sufficently well cooked. Frequently unfavorable weather or inexperienced burning caused the whole kilnful to come out worthless cinders or raw stone. At best the whole product of a kiln for a week was not more than twenty-five barrels."
10

R 20 M. B. WAKEFIELD

Nigger Lock #13 looking southeast toward High Falls. Lock #12, the third of the three locks on the bank opposite High Falls has been filled in.

Such was the improvement in making cement since that early day that, while the production was then not more than 25 barrels a week and often much less, by 1880, 550 or 600 barrels were obtained from one kiln in the same period of time.

R 21 M. B. WAKEFIELD

Nigger Lock #14, looking northeast just above the site of the High Falls Aqueduct. Note the long stay bolts that secured the wood lining to the lock wall.

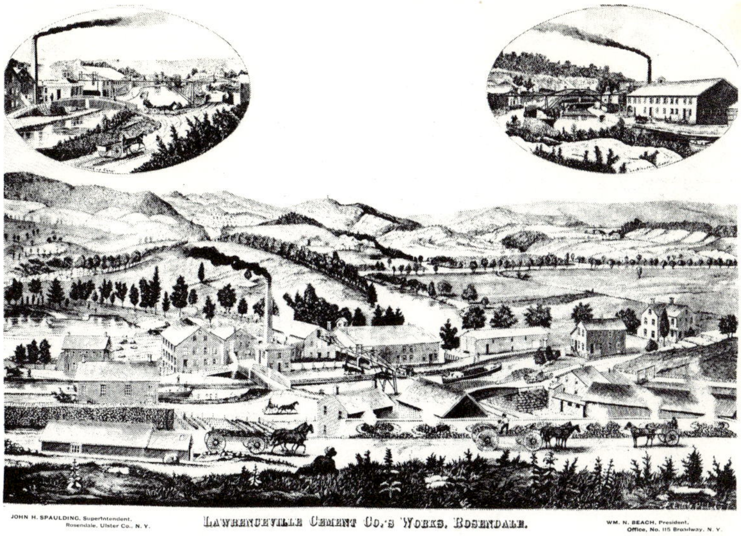

R 23 Collection of JIM SHAUGHNESSY

Lithographic overview of the Lawrenceville Cement Works. In the center, at the base of the chimney, is the square office building whose life span is traced through pictures R 24 and R 25. Note the railroad bridge used to carry cars of cement across the canal.

Mr. Lawrence later failed in business, but, determined to redeem himself, he established another mill about 100 rods westward up the canal from his first cement mill. In 1854 he had four good-sized drawkilns built, from which a railroad track was constructed high over the main road and canal into his grist mill works which had been converted into a cement mill. R 23, R 24 and R 25

R 22 M. B. WAKEFIELD

Cement kilns at Lawrenceville with Lock #9 at the left. Boats were loaded directly in front of the kilns.

Ruins of this great pioneer industry may be found bordering the canal near Lock #9 and elsewhere in the Lawrenceville-Rosendale area. R 22

For about five years after Captain Hensberger's last trip to tidewater, the cement boats continued to move down what was known as a *transfer canal.* DuBois Weber remembered when he "hired out" with his black bob-tailed team " . . . and rode up the towpath through the cement area between Rock Lock and Lawrenceville looking for a loaded boat ready to go. When you found one, the boatman would throw the towline and down you would go to Eddyville. I can remember, as if it were yesterday, taking out the last loaded cement boat."

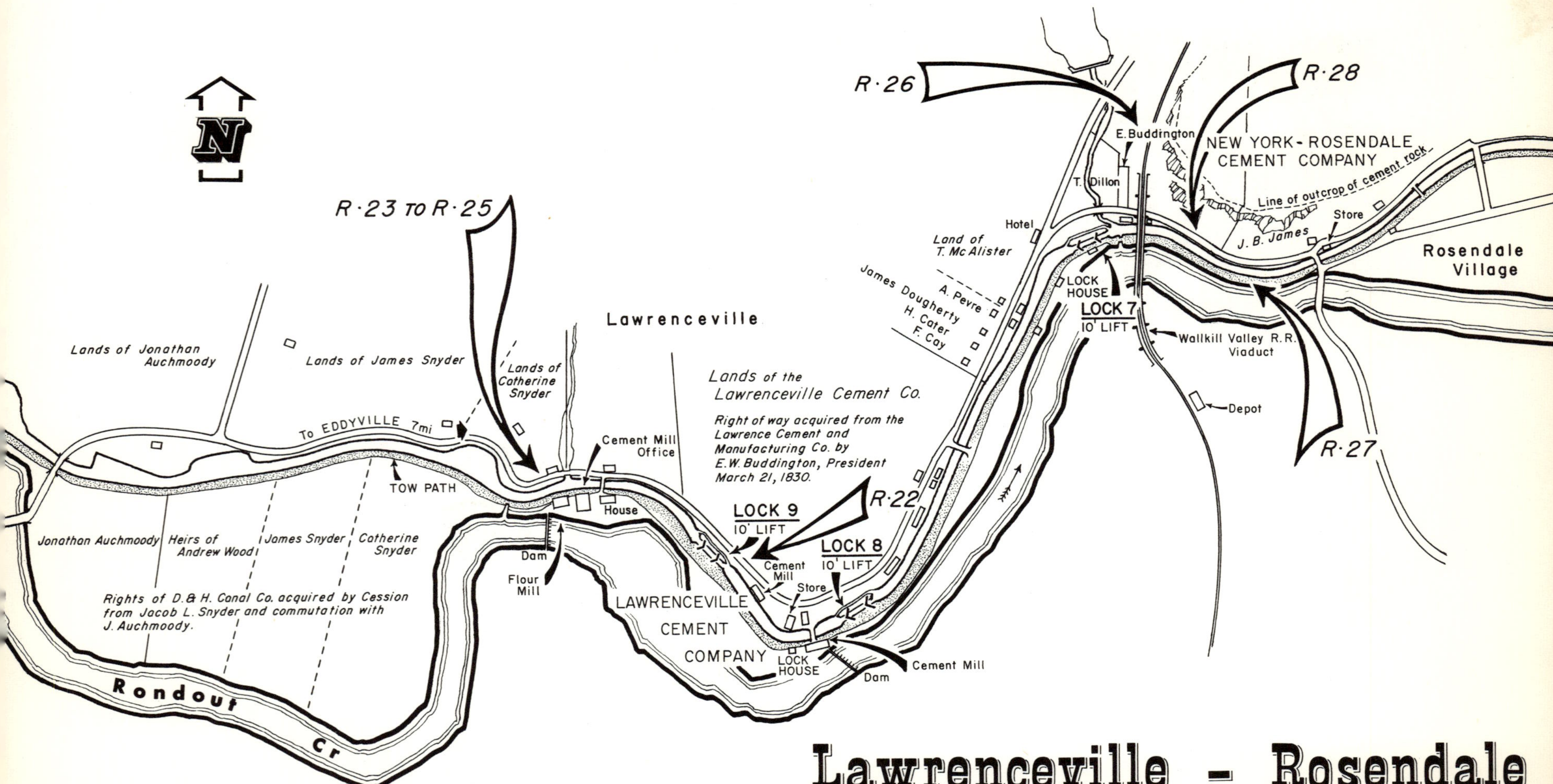

Lawrenceville - Rosendale

The Lawrenceville Cement Works office building shortly before its demolition.

R 25 M. B. WAKEFIELD

The D. and H. Canal and wagon bridge over the boat slip at the old Lawrenceville Cement Mill shown on the preceding page in a lithograph. The burned out building was at one time a flour mill which was converted to the manufacture of cement.

R 24 Courtesy ELLENVILLE PUBLIC LIBRARY

Just around a sharp bend in the river, the 140-foot-high Wallkill Valley Railroad trestle greets the eye to mark the location of the village of Rosendale. The trestle, some 900 feet in length, was completed in 1876. R 26

Rosendale in 1872 contained three churches, two hotels, two stores, a school, blacksmith shops, three wagon shops, an undertaker, a harness shop, two milliners and about 550 inhabitants. R 27, R 28 and R 29. In the 1872 Ulster County Directory, two boatmen listed Rosendale as their home, Levi Deyo and Wesley Rugar. Warren K. Atkinson, supt. foreman of the first division of the canal also made Rosendale his base of operations. The large New

York and Rosendale Cement Company obtained their cement rock from the mountain that towers above Rosendale and the Wallkill Valley Railroad trestle. R 27 and R 29

Two other cement works were found at LeFever Falls about one mile north of Rosendale. R 30. In 1873 the New York Cement Company, located at the falls, manufactured about 500 barrels of cement per day for a total of about 100,000 barrels during the season. Just below Lock #6 at LeFever Falls, known to DuBois Weber as *Rock Lock* and also *Webster's,* was

R 27 Courtesy of ELLENVILLE PUBLIC LIBRARY

A heavily loaded cement boat moving through Rosendale under the Wallkill Valley Railroad trestle. The mountain at the right contained the quarries of the New York & Rosendale Cement Company.

R 26 From original glass plate negative, Collection of GERALD M. BEST

Erie Railway train on the Wallkill Valley Railroad Viaduct at Rosendale at the time of its completion in 1876. At that time the Wallkill Valley Railroad did not own any locomotives. The engine is #208, built by the New Jersey Locomotive Works in 1854. The D. and H. Canal may be seen under the bridge just below Lock #7.

R 29 Courtesy ELLENVILLE PUBLIC LIBRARY

The New York and Rosendale Lime and Cement Company Works beside the canal as shown on a contemporary lithograph print.

located the imposing Martin and Clearwater Cement Works. Situated next to Hardenburg's Basin, the works had a capacity for producing 80,000 barrels per season and gave employment to about 100 men.

The community of LeFever Falls at one time contained a store, three hotels and about thirty dwellings. Two community leaders were John King who operated a canal grocery and Jacob Stall who was the proprietor of the Rock Lock Saloon.

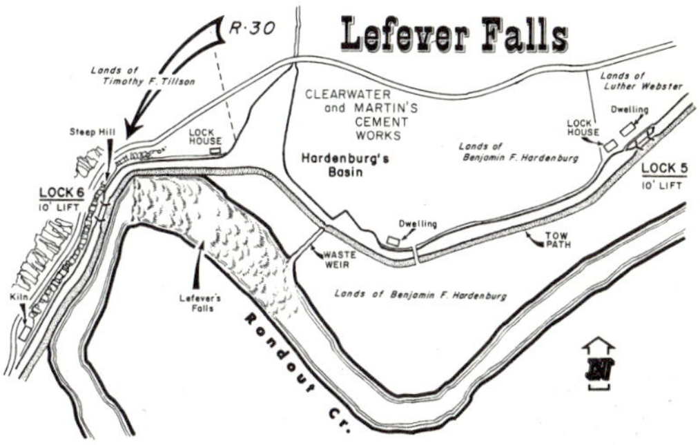

R 28 Courtesy ELLENVILLE PUBLIC LIBRARY

Looking toward Rosendale from under the Wallkill Valley Railroad Viaduct. The cement boat, riding light, is waiting to pull up to a cement dock for loading.

As late as 1956 the lock tender's bridge was still in place over the lower gate at Lock #6. Today the lock is gone, bulldozed away by a powerline company.

At Lock #4, the first of three locks known as the *Creek Locks,* is located an extremely well preserved drydock directly along the edge of the road. R 31. Further on at Creek Lock #2, canal boats entered the Rondout Creek for slack-water navigation to Eddyville. R 32. Although the community of Creek Locks in 1873 contained only two stores, a grocery, a school and about twenty-five dwellings, it did boast the busy Hudson River Cement Works. Operated by E. M. Brigham the company had their main buildings at East Kingston and their quarry at Creek Locks where they

R 31 M. B. WAKEFIELD

The well preserved drydock at Creek Lock #4. Lock #4 can be seen at the extreme right side above the wall of the dry dock. This dock is not shown on the 1854 maps of the D. and H. Canal Co., indicating that it was probably built during the Civil War period, when boat repair facilities were in great demand.

R 30 Courtesy of ELLENVILLE PUBLIC LIBRARY

Lock #6 at LeFever Falls was known to many as 'Rock Lock'. Below the lock was located the large Clearwater and Martin's Cement Works at Hardenburg's Basin. Note the 'whirling posts' on the towpath in the foreground.

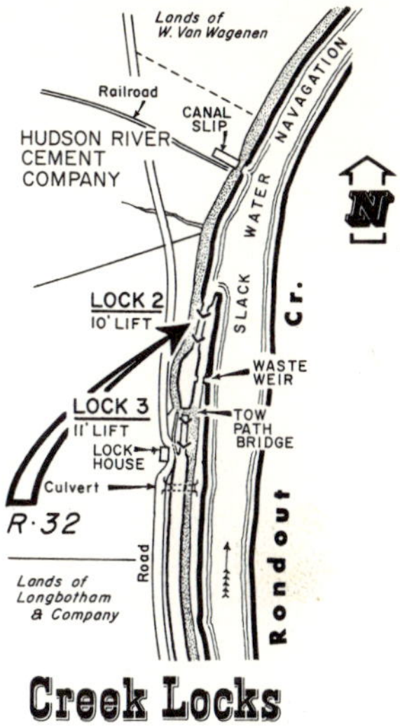

R 32 M. B. WAKEFIELD

Lock #2, the first of three locks known as the Creek Locks. From this point canal boats became stream borne on the Rondout for the two-mile run to Eddyville.

gave employment to about 50 men. The stone was conveyed from the quarry about 3/8 of a mile on an inclined railway, the descending cars drawing up the ascending ones back and forth to a canal slip just below Lock #2.

The canal company had a towpath along the north bank of the river up to the Guard Lock at Eddyville. R 36. Today the towpath has virtually disappeared due to high water and weathering along the creek. The distance from Creek Locks to Eddyville is about two miles with a treacherous point half-way down known as the *Narrows.*

The hazards of the Narrows were compounded by a rock the canallers called the *Hog's Back.* John A. McGinnis relates an accident that happened at this rock.

"My brother's wife Ann was on the boat when we locked out of Creek Locks into the river. There was a freshet and when we reached the Narrows, that's where the shoreline comes in close with a rock located about fifteen feet out from the towpath. That was the rock we called the Hog's Back. When the water was low you could see it and could drive a boat past the side."

It should be noted here that the boats passed down the river with about 150 feet of towline out. Since normal towing lines never exceeded 100 feet in length, the extra long *Neversink Locks Lines* had to be used on the semi-slack Rondout Creek waters. One cardinal rule to be remembered when going down the Narrows was to keep the boat going as fast as the water; otherwise the boat would turn around.

Mr. McGinnis continues, "On the creek you had to drive the mules harder in order to keep the boat moving faster than the water for proper steering. This feller came down behind us and of course his team wasn't towing the boat and the boat swung around and he came up on the Hog's Back and broke right in two. She was loaded with about 140 tons of coal and one part slid off and went down and the other just laid up there on the rock. People came from all over to see the wreck."

Another incident related by John McGinnis concerned a loaded boat captained by a man named Mulligan. "There was this light boat coming up by the name of *Ferguson* and it was towing against the water from a line to the middle of the boat instead of the bow timberhead. She came into the eddy near the point before the Narrows and the current hit her and she shot right across the stream. Mulligan's loaded boat coming down was hit just at the bend in the bow. A man in a rowboat nearby rowed over and got Mrs. Mulligan and all the crew off just as the boat went under. I should judge the boat went a hundred feet downstream before she snagged the bottom."

Later the company pumped the water out until it floated, then a canvas tarpaulin was wrapped around the bow and the boat floated to Rondout where the coal was removed and the boat put in drydock. "Nine of the bow timbers, which were six inches apart, were smashed right in." [11]

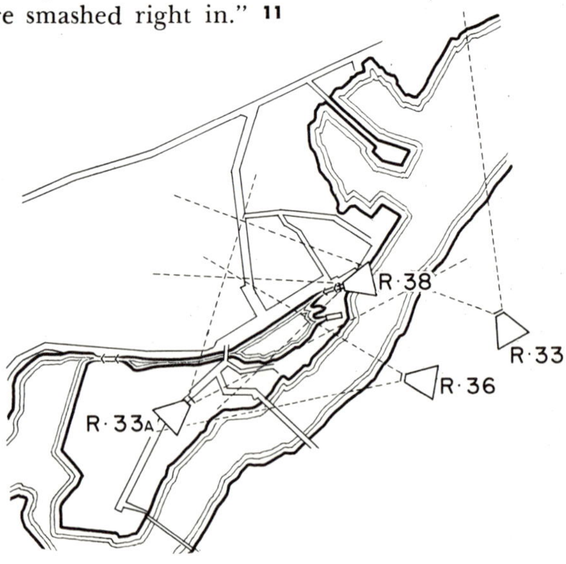

This simplified map of Eddyville denotes camera locations for the four following overviews.

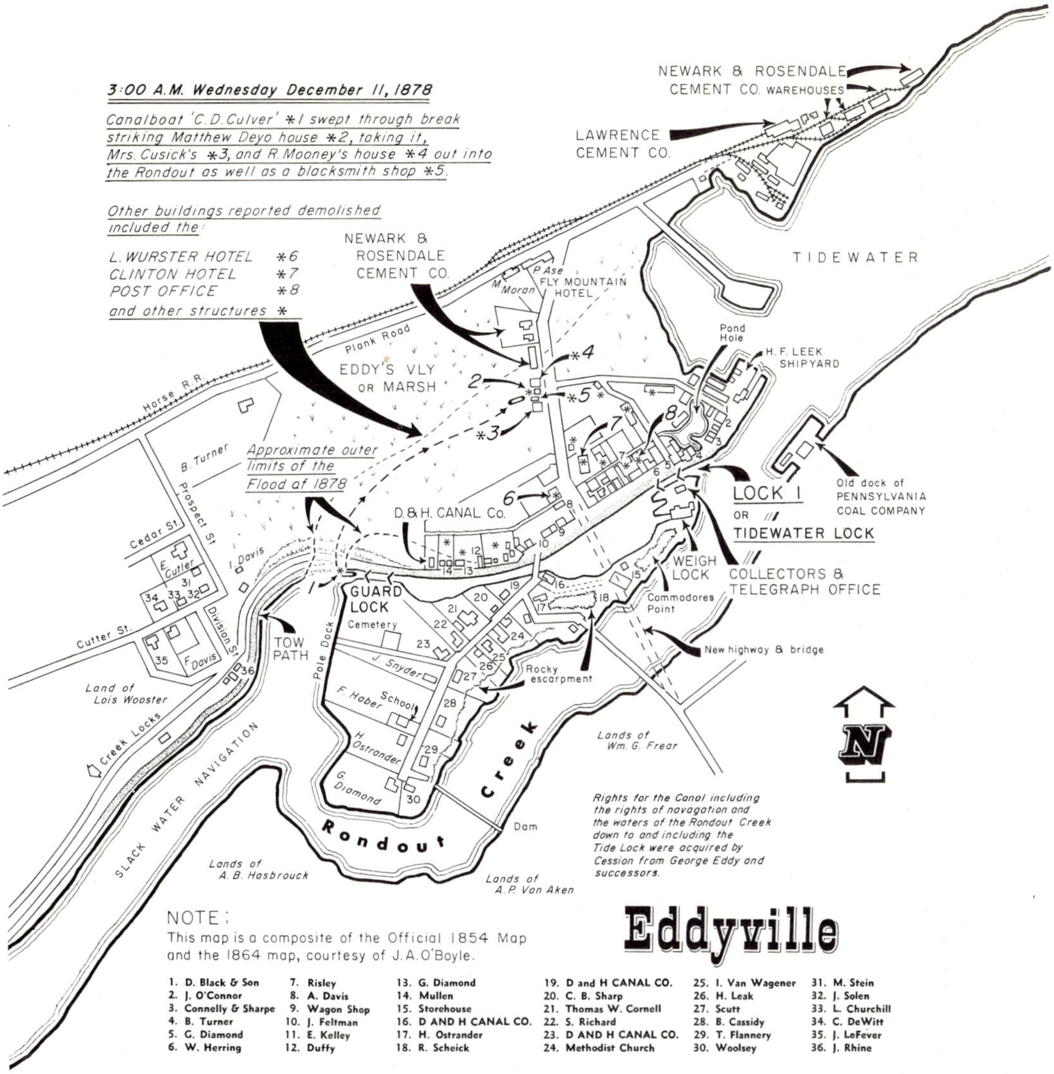

At Eddyville boats locked through the guard lock into a basin at the end of which was located the *Tidewater Lock,* or Lock #1 and the Weigh Lock. R 33 to R 39B

Eddyville was named after George Eddy, who by 1820 was well established on the west bank of the Rondout Creek, diverting water from the creek to operate a large cotton mill and saw mill. With the coming of the canal Eddyville became the head of navigation and a scene of bustling activity. The settlement spread out along the towpath and rapidly acquired the lusty waterfront atmosphere it would display for seventy-five years. Stores carrying a variety of goods sprang up and a seemingly endless row of saloons beckoned to the thirsty canallers at the end of a run. R 33A and R 38

R 33 Collection of WAYNE COUNTY HISTORICAL SOCIETY, courtesy of TOM McGINNIS

The Rondout Creek waterfront at Eddyville just below the Tidewater Lock (left). D. Black and Son and J. C. Connor used the third-floor bridge device for unloading consignments directly from shipboard. To the right of Black's is the old Leek's Shipyard while further down are the warehouses of the Newark & Rosendale Cement Company. Three cement cars on the old horse railway that ran from the waterfront to the quarry may be seen in the background (ARROW).

"At either Connolly and Schaffer's, at Black Brothers', (R 33) or at Ben Turner's Emporium, standing near the entrance to the Tidewater Lock, the canaller could find work clothes and Sunday apparel for every member of his household. The ladies could delight their hearts with figured jaconet, mull and Sarcenet muslins, Barcelona and Canton Handkerchiefs, and white cashmere shawls with merino borders. On the long shelves behind the heavy mahogany counters could be seen row upon row of tins where spices, cassia, coffee, pearlash, candles, plug paper and ladies' twist tobacco, and Macoboy Rappee snuff were kept. Close by were the half opened barrels of salt mackerel, salt beef and herring, and large casks of pungent cheshire cheese. Toward the back of the store huge coils of rope stood in front of nests of shovels, pickaxes, and cast iron pots and pans." **12**

In 1873, in addition to the canal facilities, Eddyville contained a Methodist Church, twelve stores, a hotel, the Lawrenceville Cement Factory, Newark and Rosendale Cement works and about 400 inhabitants. R 36

It wasn't until 1835 or '36 that a road was built along the creek connecting Eddyville to Rondout. Prior to that time a person was compelled to go to Kingston, thence nearly to Greenkill, where the mountain was crossed, and to Eddyville on the towpath.

There were three shipyards in the area busily engaged in building and repairing boats. In 1864 the H. F. Leek yards, across the way from Black's store, was the most important one, with some of the finest craft on the Hudson having slid down its ways. The large tug *Resolute* and the passenger steamer *A. E. Anderson,* as well as many others, floated out into the Rondout through three different managerial periods. R 33

John Snyder . . . took over from Leek, after which the business passed to John J. Baisden in 1885, who operated it until 1908.

Just below the shipyards were the sprawling warehouses and barrel factory of the Lawrence Cement Company, where three-masted schooners brought loads of staves from Maine and took on cargoes of coal or cement for the return trip. Down the creek from Eddyville, the Newark and Rosendale Lime and Cement Works maintained large warehouses next to the Lawrence Cement Company Works. The stone was quarried at Whiteport where 800 barrels a day were turned out and transported to Eddyville on a horse railroad. R 33. The quarries were entered by three tunnels, two of which were 200 feet in length and one of 70 feet. They had a perpendicular depth of about 120 to 140 feet. The company operated 17 kilns, 15 of which were in constant use (1872), and ground the rock in a mill containing 12 runs of three-foot stones.

Following the immediate post-Civil War period, the gradual decline in the shipping of coal over the canal began to have its effect on the economy of Eddyville. In 1878 a long and profitable boating season ended in disaster for the canallers and the people of the village.

Eddyville in the 1880's showing the canalfront business district on the left, and the Weigh Lock on the right with a boat poised in the entrance. In center is the Tidewater Lock #1 and to the right the Collector's Office. Tied up just before Lock #1 is a cement boat, discernible by its rather distinctive bow.

R 33A

Collection of JIM SHAUGHNESSY

The boating season lasted into the first week of December with spring-like temperatures permitting more and more boats to make it to tidewater.

It was not to last.

Fast scudding clouds moved in on warm winds on December 7th and by the 8th a warm rain had set in. It continued through the night and increased in intensity the next day. By the morning of December 10th the wild grey clouds were being pushed along by gale-force winds which transformed the Rondout into a seething froth of whitecaps. Old-timers who had seen all kinds of weather sweep up the Rondout shook their heads while they secured boats and rigging against the screeching wind.

R 35 Courtesy of ELLENVILLE PUBLIC LIBRARY

An early view of Tidewater Lock #1 and lower gate house at Eddyville. The steps on the right side were for convenience in boarding boats when the lower gates were closed and the lock filled with water.

R 34 Courtesy of PENNSYLVANIA HISTORICAL AND MUSEUM COMM., HARRISBURG, PA.

The lower gates of the Tidewater Lock in the open position to permit the passage of a cement boat out into the Rondout. The Collector's house and weigh lock are to the right out of the picture. In this later view the boarding steps have disappeared and a small addition has been tacked on the right side of the gate house. The overall rundown condition of the lower gate house would indicate the last few years of negligence preceding the final closing.

R 36 Collection of WAYNE COUNTY HISTORICAL SOCIETY, Courtesy of TOM McGINNIS

By 1897 more and more idle canalboats were being moored at Eddyville. To the left is the old iron bridge and on the hill, the Methodist Church. The canal passes under the bridge in the center for its tree-shaded approach to the guard lock. Eddy's Vly or Marsh, in the center, was the scene of great flooding and destruction in 1878 when the towpath went out at the guard lock.

As the day wore on, the creek rose higher and higher, and the business section of the village facing the weigh lock basin was looking more and more like a ghost town. Just before a premature darkness settled down, villagers looking out at the swirling creek and dam could see large trees, barns and outbuildings poise on the brink of the dam only to crash over in a splintering roar of shattering wood and wind-blown spume.

At the guard lock, men in shiny oilskins bent to the wind feverishly placing bales of hay and bags of cement on the embankment to counter the steadily rising waters. When the embankment by the guard lock went out about three o'clock Wednesday morning, people watching from the bluffs overlooking the frightful scene could not hear it over the banshee-like winds. It was only when the eye of the storm passed over, permitting the moon to shine through, that they realized the worst had happened. While they watched, the breach continued to wash away, until in about three-quarters of an hour the ominous void stretched for three hundred yards.

"Connelly and Shaffer's boat, *C. D. Culver,* which lay at the mouth of the guard lock, was the first to pass out through the break and in passing struck the dwelling of Matthew Deyo, which together with Mrs. Cusick's and Richard Mooney's houses, was lifted from its foundation and all three carried away. Matthew Deyo, his son Byron, and William Crawford were in the house when the flood came, but escaped in a small boat. Having no oars, they drifted along

R 37 M. B. WAKEFIELD

The remains of the Weigh Lock at Eddyville contains a garage at one end. It was originally covered by a substantial building as shown in picture. R **33A**

R 38 Courtesy of ELLENVILLE PUBLIC LIBRARY

The Eddyville Main Street facing the canal as seen from the lower gate bridge of the Tidewater Lock. All the structures on the right are gone, while the new bridge approach passes through the approximate site of the hotel with the second-floor porch. The weigh lock is to the left, out of the picture, from which the loaded boat entering the lock has just left.

ahead of floating houses and barns, striking the bank opposite Fitch's stone yard, when, grasping hold of trees, they succeeded in getting ashore. Beside the afore mentioned structures, the following buildings were wrecked or swept away: five tenement houses, one containing ten families, five private dwellings, two stores, the post office with its mail sacks, two hotels, twenty barns and many other buildings. The number of horses and mules and other livestock lost is terrific." 13

"It is supposed that many lives were lost, since many men, women and children on the boats were seen and heard frantically appealing for help that could not be given. At Eddyville a barn containing forty horses and mules, the property of canalmen, was taken away bodily in the flood and all the animals but one drowned." 14

R 39 M. B. WAKEFIELD

The Rondout Creek end of the Tidewater Lock #1 showing the only lock gate still in existence on the Delaware and Hudson Canal.

R 39A M. B. WAKEFIELD

The large skeletal beams of the upper drop gate still lie in place in the canal bottom at the Tidewater Lock. The metal strap hinge from which the lock pivoted shows rather dimly in the shadows along the upper end of the beams.

The towpath was rebuilt and a new guard lock erected at great cost, but the disaster left its unmistakable mark on the spirits of the people.

What might be considered the *coup de grace* for Eddyville occurred on the night before St. Patrick's Day in 1892, when a fire of mysterious origin erupted in a building on the waterfront and quickly spread on shifting winds to barns, stores and warehouses. Most residents just stood helplessly by and watched. When Captain Hensberger and his 1107 finally did arrive, six years later, it was little more than an anti-climax.

Today the weighlock has a garage in one end. R 37. Miraculously the Tidewater Lock can still boast one of its gates due in no small part to the vigilance of the owner of the nearby house. R 39. Even the skeletal remains of the upper drop gate may be seen at low tide. R 39A

Fire, decay and the building of a new bridge over the Rondout has eradicated all but three of the string of canal-side buildings (shown in picture R 38) that gave Eddyville the heady atmosphere of three-quarters of a century of prosperity.

R 39B M. B. WAKEFIELD

Lock #1 at Tidewater as it appears today. From here the loaded coal boats were tied together in groups for towing by the "Pittston" down the Rondout to Island Dock. The only surviving lock gate on the canal is located in the recess just beyond the tender's bridge.

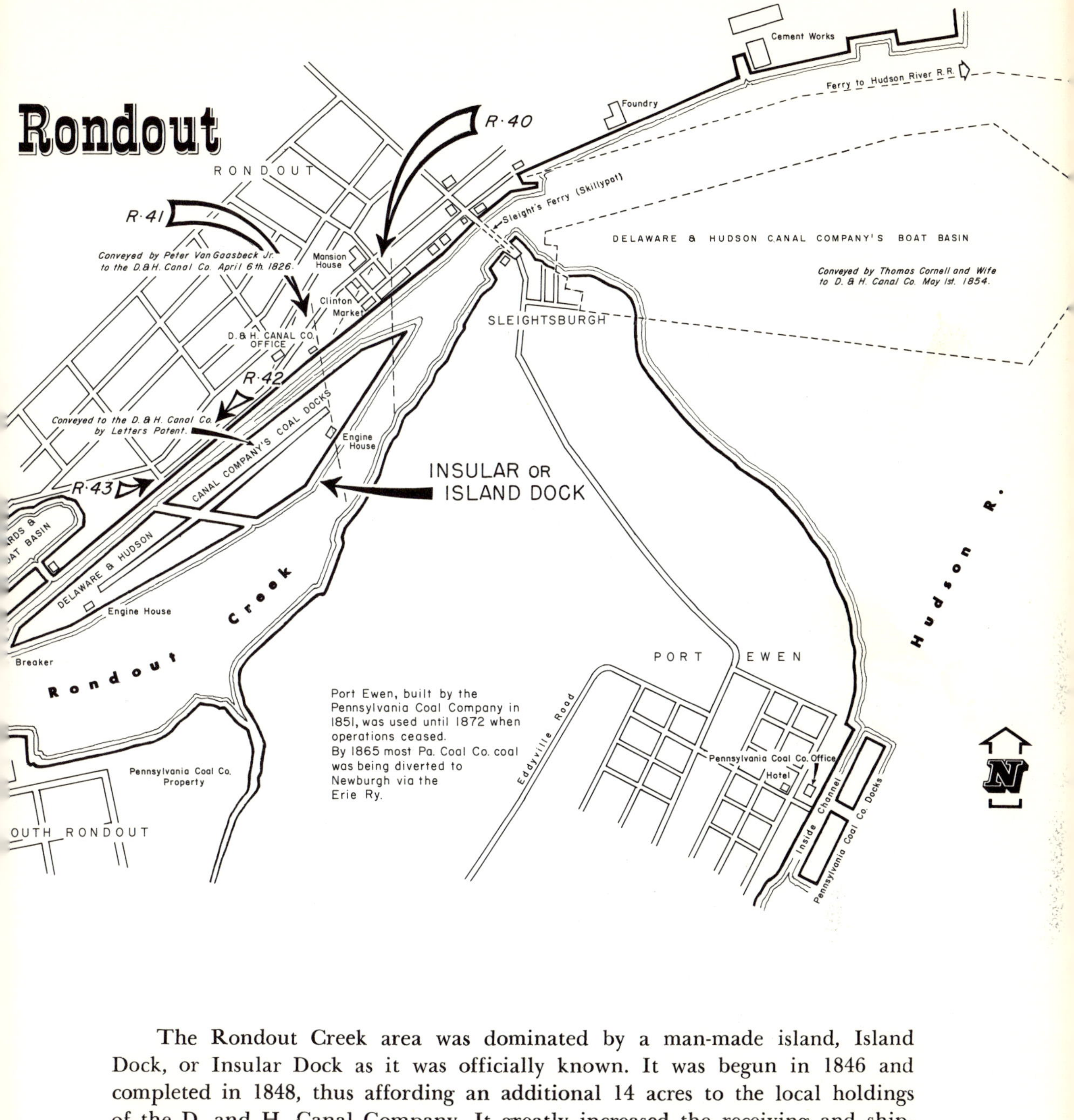

The Rondout Creek area was dominated by a man-made island, Island Dock, or Insular Dock as it was officially known. It was begun in 1846 and completed in 1848, thus affording an additional 14 acres to the local holdings of the D. and H. Canal Company. It greatly increased the receiving and shipping facilities of anthracite at the Rondout Docks. R 40, R 41, R 42 and R 43

DuBois Weber told of "the early unloading of coal by the use of big iron tubs and swarms of laborers. Toward the end, coal elevators took the coal out on one side and deposited it down through a chute on the opposite side to boats for shipment to various points. These elevators were operated by a steam boiler and would fill the coast-wise ships with the loads of eight or ten canal-boats." R 40

Summer

R 40 Collection of JIM SHAUGHNESSY

Looking west up the Rondout at Island Dock. On the left a coast-wise coal barge is being loaded from an elevator. Labor troubles arose on the waterfront when these automatic coal loading devices were first introduced.

Winter

R 41 Collection of JIM SHAUGHNESSY

The towboat "Austin", locked in Rondout ice with her brood of snow-mantled canalboats.

Upstream

R 42 Collection of GERALD M. BEST

Coalboats tied up along Island Dock at Rondout in 1898. Note the boat in the middle with the cook stove out in front of the coal bunker and the wash strung out to dry in a coaldust-laden atmosphere.

Downstream

R 43 Collection of M. B. WAKEFIELD

Looking downstream toward the Hudson River and the location of the little chain ferryboat known affectionately as the "Skillypot". It plied the creek between Rondout and Sleightsburgh. Building on the far side of the dock is the Engine House.

It was nearby that the canallers stored their boats over the winter. DuBois Weber relates the procedure that commenced about the first of April when it came time to get the boats ready.

"Early in the spring the company would notify you to get your boat ready. The boats were on the flats down toward Port Ewen (pronounced Port *June* by most old-timers) where there was very little water when the tide went out. They had a bulkhead of little boats around these flats for protection, and at hightide the boats were floated off and towed by a tug to the steamboat dock. Then you ordered your fit-out for the boat; you got the lines you needed, ordered your provisions and oats and anything for the cabin you might need. When you were ready you were put in a tow. There were two in the morning and two in the afternoon that left for Eddyville with 15 or 20 boats in a tow. At Eddyville you put out your lines and waited your turn to enter the Tidewater Lock. The boats were towed by the famous side-wheeler towboat, *Pittston,* commanded by Captain George DuBois, and piloted by North Everts." [15]

This busy valley frequently played host to epidemics. Due to the primitive sanitary facilities of the time, cholera and yellow fever often got a foothold in the settlements up and down the Rondout.

Old records reveal a West Indies schooner brought yellow fever to Rondout and environs in 1844 and took a heavy toll of life. Cholera epidemics swept through the valley in 1832, 1849 and 1852 with the 1849 epidemic taking the highest toll of life. The epidemics followed down the canal infecting the countryside.

The Delaware and Hudson Canal Company, as well as building and repairing canalboats at Rondout, operated several sea-going barges of 500 to 600 tons capacity, namely, *Rondout, Stattsburgh, Troy, Honesdale, Carbondale, Scranton, Binghamton* and others. They also operated a small tug named *Rondout,* constructed small so it could enter the canal when necessary. This sturdy craft was under the command of Captain Frank Stevens.

The Staples Transportation Company of Fall River, Massachusetts, operated a fleet of coal barges and two tugs, the *George W. Dean* and the *Cora L. Staples* which towed the loaded coal barges from the Rondout docks to eastern coastal ports. [16]

End of Run

All through the 1870's the indications of trouble ahead for the boatmen occurred with alarming frequency. The costs of running the boats and feeding the mules continued to climb far above the continuing starvation rate of seventy cents a ton. When the 1878 season opened on April 11th, eight years later, the rate was still seventy cents.

The loss of more and more business to the railroad, which could ship coal in the dead of winter, brought the graph on the tonnage charts down farther each year. As early as May of 1881, the canal company managers publicly announced that the abandonment of the canal was only a question of time, stressing the point that coal could be freighted so much more cheaply by rail that boats must be abandoned.

Droughts and the resultant low water prompted the Hon. Thomas Cornell in May of 1881 to initiate consideration for the passage of legislation "authorizing any company owning or operating a canal to construct and operate a railroad along it or in its place." [1] By 1884 the condition and appearance of the horses and mules made it abundantly clear that hard times were stalking the boatmen at every bend in the canal. Pressure was being exerted on the Society For The Prevention of Cruelty to Animals to establish a check point at some spot on the canal to compel boatmen to purchase new horses and mules. The canaller's ragged appearance belied this possibility.

The average number of trips per man in both 1884 and '85 was about ten, when thirteen or fifteen had been the average in other years. Few of the boatmen made money and if an accident was met with, such as damage to a boat or the loss of a horse, it was castastrophic.

By August of 1891 the season was so unprofitable, due to the infrequencies of trips, that the distress of the boatmen prompted the formation of a committee who went to New York in their behalf and presented the matter before

Treasurer Hartt. The proposition, as reported by the Port Jervis *Evening Gazette* of August 22, 1891, was . . . "that they receive $2.00 per day after the first two days detention at Rondout or Honesdale, or that instead of taking the usual installment of $15.00 per trip on the boats, only $5.00 be taken during the remainder of the present season, or that at the expiration of the present season the company grant each boatman a loan of $50.00, the said amount to be charged back on the payment of boats. Mr. Hartt told the committee that in two or three days the company would render its decision." There seems to be no record of a final determination.

The boating season of 1894 started off badly, with the company notifying boatmen that 150 new boats would be laid off in March. This was a severe blow to owners as they had wintered their horses expecting to boat the following season. During the last season when the boats were laid off in the middle of the year the boatmen received $100, but not so in 1894. On April 5, 1894, only twenty boats opened the season with little expectation that any of those taken off would be put back on.

An item appeared in the *New York Tribune* of June 21, 1895, stating that "A rumor is current among the Delaware and Hudson Canal Company's employees that the company has decided to abandon the canal east of Summitville and that coal from Honesdale, after being carried in boats as far as Summitville, will be placed on Ontario and Western cars at a point midway between Summitville and Phillipsport, at Rose's Lock."

EOR 1 Courtesy of HENRY G. SOETE

The last boat to leave Honesdale, #1107, under the command of Captain Hensberger shown at the tiller. Henry Soete as a boy stands on the cabin roof with his father Francis X. Soete to the left.

Some idea of the drop in business can be gauged when it is realized that in 1890 there were 800 D. and H. boats on the canal, but by 1898 only 387 were still in operation. Finally, on November 5, 1898, Boat #1107 pulled out of Honesdale with the last load of coal under the command of Captain Hensberger. EOR 1.

In 1899 the company, desiring to keep its franchise alive as far as the Mongaup Aqueduct, ran a boat up to Mongaup Basin to pick up a load of cordwood.[2] "It is stated that Charac J. VanInwegen, on the last boat that passed through the canal locks, freighted the last cargo on the canal to tidewater, being of cordwood designed for brick manufacture on the Hudson River".[3]

EOR 2 Collection of GERALD M. BEST

Bridge #1 from West's Basin at Sparrowbush showing one of the boats anchored in the canal to keep the Erie from dumping debris into the waterway. The only activity on the towpath when this picture was taken was the frequent passing of Mr. Darragh on his bicycle.

After the passing of Boat 1107 the Erie R. R. started the insidious practice of dumping rock, gravel, dirt and refuse into the canal at their Lackawaxen crossing. In an effort to thwart dumping at other points, the canal company placed boats under the two Erie bridges downstream and hired a watchman named William Darragh to patrol the right-of-way.

"There were two bridges where the Erie crossed the canal between Bolton Basin and Port Jervis, Bridge #2 on the edge of Bolton Basin and Bridge #1 where it recrossed the canal at Sparrowbush. (See BOLTON BASIN Map.) The canal Company anchored a boat under each bridge to discourage any dumping move by the Erie. My job was to patrol the canal on a bicycle from Butler's Lock to the Neversink Locks. If I saw any activity with dump cars loaded with cinders or gravel on the main line or in the yards at Port Jervis, I was to call Superintendent Rose[4] in Kingston and a restraining order would then be issued."[5] EOR 2 and EOR 3.

Mr. Darragh, still hale and hardy in April of 1965, related how he used to crack the paddle gates of Butler's Lock which "started a swell of water which flushed out the muskrats as it moved down the canal. Since I generally carried my gun with me I'd jump off the bicycle and pick em' off."

EOR 3 Collection of GERALD M. BEST

Bridge #2 over the canal looking toward Bolton Basin. The Sawmill Rift Bridge of the Erie R. R. is to the left out of the picture. In the distance just over the boat is the old hotel at Bolton Basin. The towpath from a point over the bow of the boat to the hotel marks the location of the first of the two 1885 breaks.

Photo by PAUL GERRY
William Darragh of Sparrowbush

The canal continued to operate two seasons from Ellenville, the last boat moving out in July of 1901, and five seasons from High Falls to tidewater to accommodate the cement works. [6]

All these latter-day operations were under the benevolent eye of Mr. S. D. Coykendall, the president of the Cornell Steamboat Company, who purchased all its franchises, rights and privileges for $10,000 in June of 1899. Also in 1899 the Delaware Valley and Kingston Railway Company was organized and in 1900 secured a charter. This organization was the end result of the 1881 legislative efforts by Thomas Cornell as noted earlier.

These activities were motivated by the intense desire to build a competing coal line to the Erie Railroad. Shortly the fledgling company was acquired by the Erie in connection with its purchase of capital stock of the Pennsylvania Coal Company and the Erie and Wyoming

Valley Railroad. Hence the Erie became the overseer of a ditch that was rapidly becoming a sanitation problem.

On August 9, 1899, the company issued a warning that people must cease throwing things into the canal that would pollute the water. More difficult to control was the habit of people walking and driving along the towpath.

The sanitation problem at Port Jervis reached a crisis in the summer of 1902 when much of the illness was blamed on the unhealthly condition of the old canal bed. Accordingly in November of the same year the Erie Railroad put a force of men to work digging a ditch to drain the canal of the accumulated surface waters. By May 8, 1908, health conditions were so bad along the canal that a special meeting was held and the Erie duly notified to correct the condition.

Amid all this furor rumors persisted that the canal would again reopen. In 1905 *The New York World* claimed that shipping on canals could be done at less than half the cost charged by the railroads. The story further noted that by reopening the D. and H. a saving of two dollars a ton would be realized.

March 20, 1907, saw Assemblyman Fowler introducing a bill calling for an appropriation of $1,000,000 for the reopening of the Delaware and Hudson

Courtesy of ALLEN JOHNSON

Looking west at Summitville. The touring car and improved roads are closing in on the remains of the canal.

Canal. Although all such bills were killed in the legislature, there were people who still hoped to see the days of the canal return, as evidenced in the submission of two bills by Assemblyman Cook to the legislature as late as March 12, 1912.

The last futile effort at resurrecting the D. and H. came in 1915 when an attempt was initiated in Honesdale.

The effort failed.

The days are gone when the coal boats drifted along at three miles an hour over a winding waterway through an ever-changing landscape. The only hint of this great waterway, as you pass by, is an occasional glimpse of a weed-grown lock chamber or the strange towpath mound winding its way along contour lines down the valley.

It is still possible to stand and look down through a lonely abandoned lock and out into a passing basin and almost see a light boat moving slowly toward the lower gate. The boat is a vision, but the water is real, the snubbing post is real, as are the youngsters who sometimes come up quietly behind you just as you are snubbing up, and shatter the illusion.

The vision of the boats will come and go, but the posts, water, and children will go on forever.

THE END

Photo by PAUL GERRY

Acknowledgments

In reflecting back to the germ of this book and how it was nurtured, four people come into focus; Mrs. Grace K. Bentley of the Wayne County Historical Society who showed me every courtesy in my discovery years of the D. and H. Canal, the late Mrs. Fred Terwilliger of the Minisink Valley Historical Society whose abiding interest and concern for all things historical left a lasting impression and finally the late DuBois Weber and John A. McGinnis whose stories precipitated my total surrender to the fascinating lore of the D. and H. Canal.

I am deeply indebted to Mr. Gerald M. Best who offered his extensive photograph collection for consideration. Also Mr. Charles E. Fisher, President of the Railway and Locomotive Historical Society in granting permission to quote from Society Bulletin #82.

To Jim Shaughnessy, noted railroad photographer and author for assistance and technical services in the reworking of old canal pictures for quality book reproduction. Also for making his personal photo collection available.

Special thanks to Mr. Frank Thompson, Director of the Canal Museum of New York State in Syracuse for an opinion and review of chapter content when at a critical stage.

To Dr. Irwin Richman, Chief, Section of Science, Industry and Technology of the Pennsylvania Historical and Museum Commission for editorial assistance as well as photographs and memorabilia of the Commission, Mr. James A. O'Boyle of Brooklyn for his collection of D. and H. memorabilia, notes,reminiscenses and assistance in clearing up technical data relative to locks, boats and virtually all things concerning the 'canawler'.

To J. Vance Hunt of Hawley for photographs of the canal and the Pennsylvania Coal Company Gravity and to Henry Soete, Honesdale's enthusiastic fount of information on all things canal and gravity.

Special thanks to Cornelius Cuddeback III, Calvin Crane, President of the Minisink Valley Historical Society and D. Nelson Raynor, Port Jervis Historian for opening up the treasures of the societies collection for research on numberless occasions.

To Mrs. Marion DuMond, Head Librarian of the Ellenville Public Library and her staff for kind assistances as well as the interest and help of Mrs. Phyllis Walker, Miss Katherine Terwilliger and Mr. Milton Townsend.

A most sincere thank you to Mrs. Inez Gridley, a grand lady whose encouragement and literary magic from the very beginning was the vital spark that moved the book inexorably to completion. To Mrs. Pauline Hill and Rosemarie Nitschke, sincere appreciation for perhaps the most uninspirational chore, that of proof reading and correcting galley proofs and to Mr. Michael O'Donnell for his guarded reading of the preliminary draft.

Paul Gerry, Sullivan County photographer, for processing and printing hundreds of present day photos of the canal, Malcolm Booth for data and photos of the Neversink Aqueduct and Richard Wright of the Canal Society of New York State.

To the Steingart's, Nathan, Alan and Monte, particularly Alan whose creative flair improved many a page make-up when the pressure was on. One guy who aged prematurely through all of this, sat at a linotype machine and demonstrated great fortitude and understanding . . . thanks Ross. Also a word of appreciation to Bill and Gretchen and the rest of the staff, who were exposed to this effort in one form or another.

And finally, to my wife Barbara for her patience with the sudden and unplanned trips to a library or museum to go through some old papers or a quick trip to some obscure canal ruin and the bruhaha that would erupt when a chapter didn't end on the proper page or the picture that just wouldn't fit on the correct page to collate with the text, etc. etc. . . . my sincere thanks.

Footnotes

CHAPTER 1 — The BEGINNINGS

1. The Swiss spelling uses the letter "z", Wurtz.
2. One other survey route that received considerable attention was a proposal to tunnel the Shawangunk Mountains to connect Point Carpenter with Newburgh on the Hudson River.
3. H. F. Bartle, "The Coal Canal", Pike-Wayne **Eagle,** (Hawley, Pennsylvania), June 18 - September 24, 1959.
4. Pioneer meeting place of early men of finance.
5. Malcolm A. Booth, The Delaware and Hudson Canal, with special emphasis on Deerpark, N. Y. (Typescript thesis for State University of New York College at Oneonta, N. Y., 1965), Pg. 38.
6. Bartle, op. cit., pp. 12-13.
7. Originally established in Bloomingburg(h) in 1826 as the **Whig** but moved to Monticello shortly thereafter under the banner of **Republican Watchman.**
8. The D. & H. Gravity R. R. between Honesdale and Carbondale, was opened August 8, 1829.
9. Delaware and Hudson Company, A Century of Progress, History of the Delaware and Hudson Company, 1823-1923 (Albany: J. B. Lyon Co., 1925) p. 19.

CHAPTER 2 — The LION and the GRAVITY

1. At the time of this important assignment Mr. Allen was not yet twenty-six years of age. He later went to New York, Lake Erie and Western Railway and became its president in 1843. In 1870, when nearing seventy years of age, he was appointed consulting engineer on the Brooklyn Suspension Bridge.
   Recent research by the eminent Gravity R. R. authority, Dr. Edward Steers, reveals that Allen was going to England on his own to investigate locomotive developments and was asked by the D. & H. Canal Co. to represent them. Sources show he was not directly in their employ.
2. Century of Progress, op. cit., p. 50.
3. Ibid., pp. 56-57.
4. In 1892 Mr. John Torrey of Honesdale wrote a paper describing in great detail the early years of the Gravity Railroad. The manuscript was purchased from Mr. Torrey by the Board of Managers of the D. and H. Canal Co. in the same year (1892). A copy of the manuscript was obtained by Mr. Gerald M. Best, through the efforts of Mr. Charles E. Fisher of the Railway and Locomotive Historical Society.
   Bulletin #82 (currently out of print) of the Railway and Locomotive Historical Society featured this manuscript in an article by Mr. Best. Permission has been granted by Mr. Fisher, President of the Railway and Locomotive Historical Society and the author, Mr. Best, to quote from this comprehensive report.
5. Gerald M. Best, "The Gravity Railroad of the Delaware and Hudson Canal Company" (Boston: Bulletin #82, Railway and Locomotive Historical Society, April 1951), p. 10, quoting John Torrey Manuscript.
6. Best, Torrey manuscript, op. cit., pp. 10-11.
7. Ibid., p. 11.
8. Ibid., pp. 11-12.
9. Ibid., p. 14.
10. Ibid., pp. 15-16.
11. Ibid., p. 17.
12. Best, op. cit. p. 18.
13. Ibid., p. 21.

CHAPTER 3 — The CANAL its LOCKS and BOATS

1. Century of Progress, op. cit., p. 38.
2. Dr. William L. Cuddeback, Deerpark, The Delaware and Hudson Canal, An address before the Minisink Valley Historical Society, February 22, 1928, p. 20.
3. Bartle, op. cit., In most instances locks were built to an even 15 foot width.
4. Alvin Fay Harlow, "Old Towpaths" (originally published in 1926 and reprinted in 1964 by Kennikat Press, Port Washington, N. Y.) p. 339, Permission granted May 6, 1965.
5. DuBois Weber recalled the scow brig as being a standard 140 ton boat heavily loaded in the bow with sand or cement. The driver would ram the scow into or up on top of the ice area in order to break it up.
6. Thomas J. Ham, written 1903, Wayne County **Independent,** August 3, 1954.

CHAPTER 4 — The Lackawaxen SECTION

1. T. J. Ham relates "Twenty-two years after that, March 7, 1849, the D. and H. Canal Company surveyed the two lots on the southwest corner of 3rd. (now Court) and 9th. St. comprising together 100 feet by 130 feet for $1,000; precisely the same amount Wurts paid Sam Kimble for the entire tract comprising many acres."
2. Gill's Latches: Just a short distance above Seeleyville the light and loaded tracks crossed at grade. Henry Soete of Honesdale recalls they had a primitive type of inter-locking device that consisted of a series of latches and levers, and since the nearby farm was owned by a man by the name of Gill the name Gill's Latches came into common use for the crossover point. Other sources indicate latches was another name for self-acting switches.
3. Edwin B. Callaway, "My Recollections of Gravity Days", Wayne **Independent.** Honesdale, Pa., 125th Anniversary edition, August 3, 1954.
4. Delaware and Hudson Railroad Bulletin, August 1, 1935.

5. Port Jervis **Evening Gazette,** July 8, 1869.
6. Ibid, September 11, 1869.
7. Feeder ponds used by the Delaware and Hudson Canal Co. in the Wayne County area were as follows: Millers Pond, Belmont Reservoir, Stevenson Pond, Long Pond, White Oak Pond, Elk Pond, Lower Woods Pond, Yarnell Pond and Cajaw Pond.
8. Archives: Minisink Valley Historical Society, Port Jervis.
9. Bartle, op. cit., p. 53.
10. M. J. McAndrew, History of Hawley, Pennsylvania, July, 1927, pp. 40-41.
11. McAndrew, op. cit., p. 79, James O'Connor Letter, (Description of first construction locomotive to enter Hawley.)
12. Archives: M. V. H. S., op. cit.
13. In Goshen the first telegraph key was set up in the bar room of the old Occidental Hotel for its first local public display and immediately became a source of amazement to the people.

CHAPTER 5 — DELAWARE & LACKAWAXEN AQUEDUCTS

1. Phineas G. Goodrich, "The History of Wayne County" (Honesdale: Haines & Beardsley, 1880), P. 121
2. Archives: M. V. H. S., op. cit.
3. John Willard Johnston, "Reminiscences and Descriptive Account of the Delaware Valley and Its Connections Aiming to Extend from Pond Eddy to Narrowsburg", (Port Jervis, N. Y.: Minisink Valley Historical Society, 1900) p. 37.
4. Johnston, op. cit., pp. 38-42.
5. Graveling was a process of spreading fine gravel over the upstream face of the dam so as to chink up the small leakages common to dams employing brackets.
6. Johnston, op. cit., pp. 38-42.
7. Booth, op. cit., p. 54.
8. David B. Steinman, "The Builders of the Bridge, The Story of John Roebling and His Son," (New York: Harcourt, Brace, World, Inc., 1945), p. 103.
9. Each wire in Roebling's cables were tested to twelve hundred pounds tensile strength. It was also held in a vise or in pliers and bent to a right angle, then bent over to the same angle in the opposite direction, then straightened up. If it stood this test without sign of fracture, it was accepted; if not, it was rejected.
10. Their fears were unwarrented since the weight remained constant. The boat displacing an equal weight while in the trunk of the aqueduct. Incidentially the spans are not of equal length, they vary from 132 to 142 feet.
11. The New York, Lake Erie and Western Railway existed as such until 1893 when it went into bankruptcy. After reorganization it became the Erie Railroad in 1895.
12. Large rafts that had four oars, two at each end. Singles had one oar at each end.
13. Century of Progress, op. cit., p. 139.
14. Johnston, op. cit., p. 143.
15. Further search of the Johnston Manuscripts reveals that Engineer Lord was depending on alcoholic spirits as well as pumps during the Lackawaxen abutment operations.
16. Nearpass Collection, (Port Jervis, N. Y.: Minisink Valley Historical Society, March 5, 1898)

CHAPTER 6 — The DELAWARE SECTION

1. Hamilton Child, Gazetteer and Business Directory of Sullivan County, N. Y. for 1872 - 73, (Syracuse: The Journal Office, 1872), p. 151.
2. Author of the Johnston Manuscripts.
3. Johnston, op. cit., p. 284.
4. Ibid., p. 285.
5. Ibid., p. 290.
6. Ibid., pp. 230-31.
7. Ibid., pp. 191-92
8. Ibid., pp. 149-50.
9. DuBois Weber, taped reminiscences.
10. **Tri-States Union,** April 20, 1866.
11. Port Jervis **Evening Gazette,** October 31, 1882
12. **Tri-States Union**, Port Jervis, June 6, 1862.
13. Ibid., June 6, 1862.
14. Archives: M. V. H. S., op. cit.
15. Herman R. Lytell, "The Delaware and Hudson Canal", W. P. A. Local History Project, (Port Jervis, N. Y., 1935 - 1937), pp. 108-109.
16. **Tri-States Union,** Port Jervis, August 7, 1851.
17. Lytell, op. cit., Summary of Excursion News Items.

CHAPTER 7 — The NEVERSINK SECTION

1. Port Jervis **Evening Gazette,** September 18, 1877.
2. DuBois Weber, taped reminiscences.
3. Port Jervis **Evening Gazette,** May 15, 1873.
4. Malcolm A. Booth, op. cit., p. 56.
5. Ibid, p. 58.

CHAPTER 8 — The SUMMIT LEVEL

1. Port Jervis **Evening Gazette,** September 7, 1869.
2. Boatmen residing in Westbrookville in 1873 were Joseph M. J. Snyder and Samuel Skinner.

3. Port Jervis **Evening Gazette,** October 15, 1870.

4. Johnston, op. cit., pp. 95-96.

5. Ibid., p. 105.

6. Bartle, op. cit., p. 63.

7. **Tri-States Union,** Port Jervis, July 18, 1862.

8. Johnston, op. cit., p. 203.

9. Ibid, pp. 204-205.

10. James S. McEntee cites in his "Story of the Canal" that ground breaking ceremonies took place on the 13th day of July, 1825 at Mamakating (now Wurtsboro) in a field in the rear of Dr. Morrison's House. Mrs. Thomas Moore, in 1965, notes the ceremony taking place in 1825 behind the Kip home on Grand Street.

11. John F. Crance, John Craven, Robert Dougherty, Daniel Flanigan, David S. Harding, Peter Mackin, James Malone, Patrick McCue, Bernard Traynot, Patrick McElroy, James McKenna, Patrick McNamee, James Moley, John Moley, Edward Moore, James Murphy, William L. Rose, and George H. Schoonmaker.

12. The hamlet was originally named Beatysburg by the canal company in honor of Obadiah Beaty from whom the right-of-way was purchased.

13. Port Jervis **Evening Gazette,** May 18, 1887.

CHAPTER 9 — The ELLENVILLE SECTION

1. Franklin Erath, Henry Frear, George H. Goodsir, Joseph P. Gay, Christian Heinzmann, Patrick Quigley, John Roper, Edwin Schart, Christopher Welch and Patrick Welch.

CHAPTER 10 — The RONDOUT SECTION

1. Official 1856 Delaware and Hudson Canal Company map spelling.

2. As of 1856.

3. Boatmen residing in Port Jackson (Accord) in 1873: Jonathan Bennett, Moses Brodhead, Charles Dunn, Simon J. Gillispie, Jeremiah Schoonmaker, John U. Schoonmaker and Jacob A. Smith.

4. James S. McEntee, Story of Early Canal Days, 1874.

5. In Alligerville the old canal store operated by one A. Snyder still stands. Before new siding was installed over the old weathered sides the following items were still barely legible on both sides of the front door, alspice, pepper, ginger, coperas, sulphor, (sic.), salnitre, salsoda, starch, opium, camphor, laudanum, pink root, senna and oil spike. Coal was unloaded at Alligerville for transport through the Clove Valley and up the hill to the Mohawk Mountain House.

6. Some oldtimers claim it was 1856.

7. DuBois Weber, taped reminiscences.

8. Malcolm A. Booth, op. cit., p. 60.

9. The oldest of the three coins included in the commemorative box was a 25-cent piece dated 1806, a 50-cent piece, dated 1818, and a 10-cent piece dated 1821. The largest of the silver plates, almost five inches square reads, "Erected in the year of our Lord. 1826: by the Delaware and Hudson Canal Company. John Bolton, Pres." and lists the names of eighteen managers, officers and engineers of the company. The inscription at the bottom of the plate reads: Marbletown, Sept. 2nd," and is followed by the signature of the plate maker.

   A smaller silver plate, two inches wide and almost three inches tall, is inscribed with the names of Governor DeWitt Clinton and Lieutenant Governor James Talmadge and the year, 1826. The reverse side bears the signature, I. VanSteenberg and "Mak" representing the maker.

   The third plate, circular and two and one-quarter inches in diameter bears the inscription "Mount Horeb Kingston Ulster County N York" around its perimeter. The word "Chapter" and the date "Sept. 1826" are inscribed in the center. This medallion was probably placed in the box by the Masonic chapter of that name still in existence in Kingston. The two silver rectangular plates and the three coins are on display at the New York State Senate House Museum in Kingston. The silver disk was given to the Mount Horeb chapter of the Royal Arch Masons in Kingston, the name of the organization that was inscribed on the piece. (Courtesy Central Hudson, Co.)

10. Nathaniel Bartlett Sylvester, History of Ulster County, N. Y. (Philadelphia: Everts and Peck, 1880.

11. John McGinnis, taped reminiscences.

12. Leon Sciaky, "The Rondout and Its Canal", New York History, July 1941.

13. Port Jervis **Evening Gazette,** December 17, 1878.

14. Port Jervis **Evening Gazette,** December 12, 1878.

15. The Pittston's schedule for the tows were as follows; daily except Sunday, 6 a. m., 10 a. m. and 1 p. m., 3 p. m. for light boats from the Steamboat Dock. Loaded boats towed down from Eddyville at 7 a. m., 11 a. m. and 2 p. m. and 5 p. m.

16. Data courtesy of James A. O'Boyle of Brooklyn, N. Y.

CHAPTER 11 — END of RUN

1. **Tri-States Union,** Port Jervis, May 20, 1881.

2. William Darragh, taped reminiscences.

3. Cuddeback, op. cit., p. 20.

4. J. Graham Rose, Superintendent for the Cornell Steamboat Co.

5. William Darragh, taped reminiscences.

6. James A. O'Boyle, reminiscences.

8
9
4
17
4
21

Route of the DELAWARE & HUDS
and GRAVIT

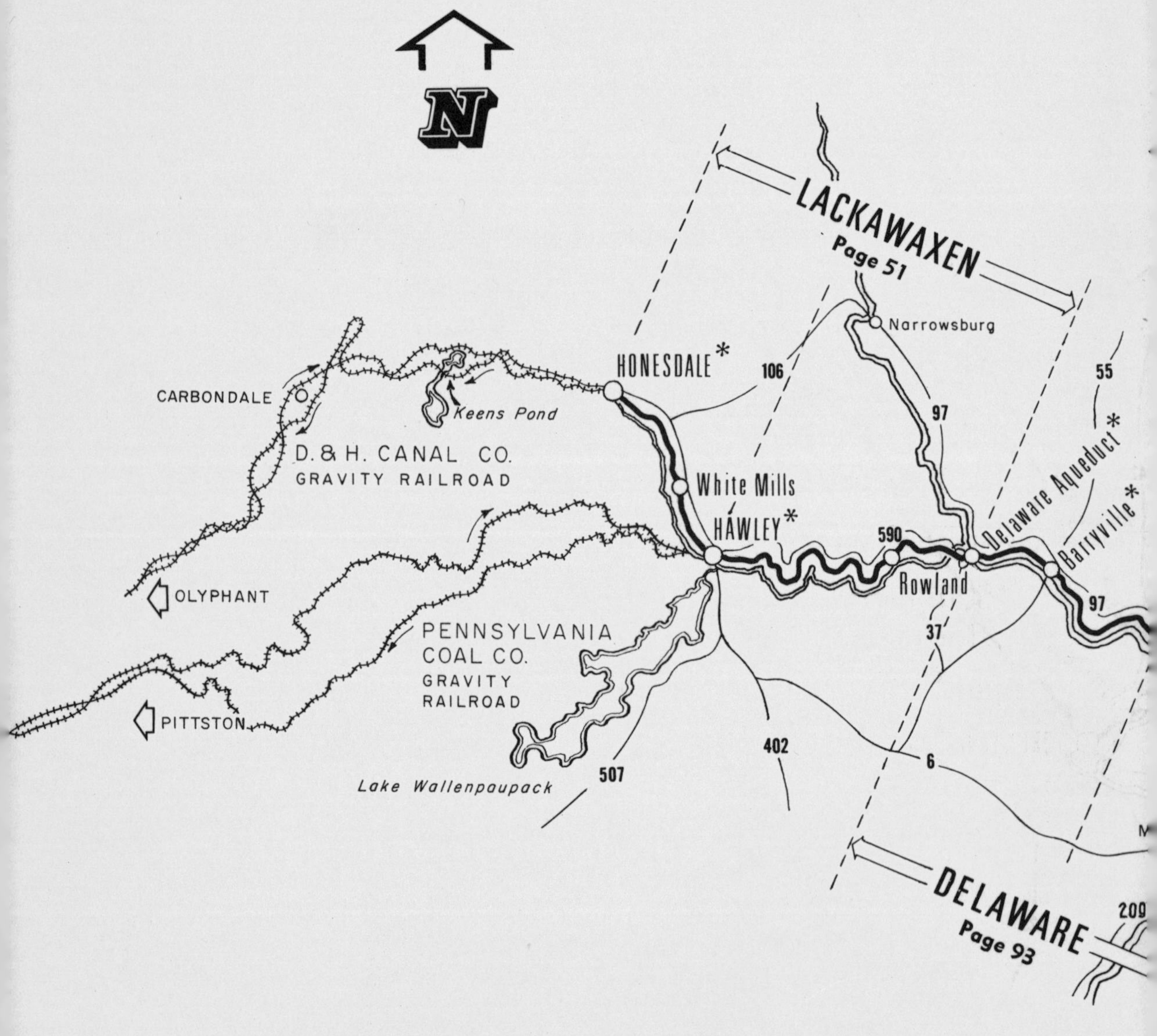